FI-SCI

THEORY AND INTERPRETATION OF NARRATIVE
James Phelan, Katra Byram, and Faye Halpern, Series Editors

FI-SCI

AVATARS OF SCIENCE IN FICTION

Rhona Trauvitch

THE OHIO STATE UNIVERSITY PRESS
COLUMBUS

Library of Congress Cataloging-in-Publication Data

Names: Trauvitch, Rhona, 1981– author

Title: Fi-sci : avatars of science in fiction / Rhona Trauvitch.

Other titles: Theory and interpretation of narrative series

Description: Columbus : The Ohio State University Press, [2026] | Series: Theory and interpretation of narrative | Includes bibliographical references and index. | Summary: "Draws on narrative and rhetorical theory to identify how mapping fiction onto science helps us conceptualize complex and even counterintuitive scientific phenomena. Analyzes fiction by Jorge Luis Borges, Julio Cortázar, Ana Menéndez, Gabriel García Márquez, and others"—Provided by publisher.

Identifiers: LCCN 2026032196 | ISBN 9780814216255 hardback | ISBN 0814216250 hardback | ISBN 9780814285213 ebook | ISBN 081428521X ebook

Subjects: LCSH: Fiction—History and criticism—Theory, etc. | Literature and science | Fiction—Technique | Fictions, Theory of | Narration (Rhetoric)

Classification: LCC PN3352.S34 T73 2026

LC record available at https://lccn.loc.gov/2026032196

Cover design by Ashley Muehlbauer
Text design by Juliet Williams
Type set in Adobe Minion Pro

♾ The paper used in this publication meets the minimum requirements of the American National Standard for Information Sciences—Permanence of Paper for Printed Library Materials. ANSI Z39.48-1992.

To Clio, Josepha, and Gabriel

&

To Jean-Philippe, Adam, and Anton

CONTENTS

ILLUSTRATIONS

TABLES

PREFACE

Flash Fi-Sci:

> The titular character of Virginia Woolf's *Jacob's Room* is never directly observed by the reader; he becomes known only through descriptions by other characters.
>
> Dark matter has not (yet) been observed directly; its existence is only known by gravitational effects on its surroundings.
>
> Aspects of Jacob are analogous to aspects of dark matter.

This Carrollesque, "Why is a raven like a writing desk?" moment is brought to you by Judith Chavarria, a student in my Spring '24 course Quantum Narratology and Other Curiosities. For her final project, she located an avatar of science (dark matter) in fiction (*Jacob's Room*).

When a concept in fiction is analogous to a concept in science, one can use the former to reflect upon the latter. I call this drawing on fiction to reason about science "fiction → science," or, "fi-sci" pattern mapping. This book is about the why and how of fiction-to-science pattern mapping. It is a call to adventure: Cosmonauts and fictionauts, beam on board to explore how the seemingly distant constellations of science and fiction can be connected by the wormhole of analogy.

INTRODUCTION

Pattern Mapping

Fabulous Fiction, Stunning Science

In the twentieth century, scientists made astonishing discoveries, such as the process of time dilation, the condition of quantum superposition, and the state of quantum entanglement. They postulated the unvisualizable, like dark matter and an eleven-dimensional spacetime. What makes these breakthroughs extraordinary is that unlike the findings of earlier periods, the phenomena just listed are difficult to conceptualize. How might we think about—or even begin to approach—science that ranges from perplexing to unobservable to entirely counterintuitive? In this book I propose that we can do so with a tool that is not an instrument like a microscope, nor a machine like a particle collider, but a rhetorical device: fictionality. More specifically, we can wield the human capacity to invent, to imagine the nonactual, as a means to the end of conceptualizing science. If you find my proposal persuasive, we'll have arrived at a remarkable conclusion: Our abilities to perceive strange and confounding science phenomena may be boosted by our experiences with fiction.

To explore this hypothesis, we set off on a path that begins with the aforementioned question's core: How do we think about—or even begin to approach—*any* new thing? In *Surfaces and Essences,* Douglas Hofstadter and Emmanuel Sander show that we do so by making analogies: We automatically compare every novel thing we come across to something we encountered earlier that is in some way similar. Consider your experience with gravity. When you see an apple rolling across the table, you instinctively know that if it reaches the edge while maintaining its speed it will fall to the floor—even if you've never found yourself in that particular circumstance before. You know it because you recognize that this situation is like your countless previous observations of gravity's pull. You have made an analogy.

If this sort of reasoning is instinctive—automatic—why not use it to conceptualize confounding science? The answer is that apparently there is nothing we can compare phenomena like quantum superposition *to.* Superposition describes the condition of a quantum particle that is simultaneously in two incompatible states. Erwin Schrödinger famously illustrated the bizarre nature of quantum superposition by positing a concurrently living-and-dead cat—which intuitively does not seem possible.[1] By definition, a concept that is counterintuitive cannot be linked to anything we would recognize from our embodied experience of the world. This is no rolling apple.

Since quantum superposition does not correspond to anything in our everyday, observed reality, to find something comparable to it—which we must do in order to make an analogy that will in turn allow us to conceptualize it—we may have to seek objects that do not exist in reality, ones unconstrained by facts, referentiality, or correspondence to embodied experience. Such objects can be found in fiction. James Phelan defines fiction as "intentionally communicated invention, projection, or other means of directing an audience to consider nonactual states or events" (Gammelgaard et al. 7). To begin reaching toward the counterintuitive, we could sure use concepts that bear these characteristics.

Let us then forage fiction for something analogous to quantum superposition. "Comparison takes place via a structure-mapping process of alignment of conceptual representations" (Gentner and Medina 267); what in fiction aligns with the conceptual representation "something is in two different states at once?" Fictional figures that are simultaneously in different states can manifest as ones whose ontological status is ambiguous. Think of characters who are both fictional and nonfictional at the same time. This situation perfectly describes the Muppets. Kermit the Frog and Miss Piggy are fictional characters in *The Great Muppet Caper* (1981) and *Muppets from Space* (1999). But when Kermit appears as a guest on Conan O'Brien's talk show ("Conan Interviews Kermit the Frog") or when Wendy Williams interviews Miss Piggy ("Wendy Williams Show"), we intuitively understand that the Muppets both retain their fictionality and present as nonfictional beings. A phenomenon in fiction thus shares a pattern with a phenomenon in science: An entity is simultaneously in two seemingly incompatible states. In their Muppetness, the

1. I introduce Schrödinger's thought experiment in more detail in chapter 5.

Muppets serve as intermediaries. Our ability to imagine *them* at least somewhat familiarizes the similarly patterned yet counterintuitive phenomenon of quantum superposition.

This cognitive progression whereby fictionality is used as a source that helps us conceptualize the target of science is valuable even when we need not depend on the former to generate an inkling about the latter. Otherwise put, cross-disciplinary analogies can be harnessed to immense cognitive benefit even when we're thinking about science of the more "user-friendly" variety, for example, as I argue, evolution by natural selection. Describing scientific models, Max Black maintains that "metaphorical thought is a distinctive mode of achieving insight." What he argues about scientific models applies to metaphors in general: They "help us notice what otherwise would be overlooked . . . to *see new connections*" (237). It follows that reasoning about any phenomenon with fictionality can afford us the capacity to perceive it in more comprehensive and novel ways. So although the original impetus for the present study was to explore a hunch that fictionality can illuminate counterintuitive science, it has grown to encompass the more capacious claim that fictionality has the potential to illuminate any kind of science.

The recognition that a phenomenon in fiction is analogous to a phenomenon in science can be described as fiction-to-science, or fiction → science pattern mapping. For ease of reference, I shall abbreviate "fiction → science" to "fi-sci."[2] When identifying fi-sci analogies, we use fictionality to better understand science, that is, we use it rhetorically, as "a means to an end" (Nielsen et al. 63). The fiction functions as a mechanism that directs an audience to consider nonactual states or events—but in the present configuration, specifically for the purpose of facilitating their comprehension of science. With this framework I thus add to narratologists' examination of what "rhetorical theorizing about fictionality can do for our understanding of literature" (Gammelgaard et al. 2) an exploration of what rhetorical theorizing about fictionality can do for our understanding of *science*. This book's mode of investigation involves searching the domain of fiction for avatars of given science patterns, that is, cross-disciplinary instantiations of these patterns. To practice pattern mapping we will ask, Can we find an avatar of this science pattern in fiction? How does this pattern manifest in fiction?

---

2. Literary scholars typically abbreviate science fiction to "sci fi" or "sf." To distinguish fiction-science from science fiction, I use the shorthand "fi-sci." But whereas "sci fi" describes a genre, whereby one term qualifies the other (i.e., a sciency-type of fiction), "fi-sci" represents a parallel between the two domains: A pattern in fiction is analogous to a pattern in science. I credit James Phelan with proposing the term *fi-sci* to characterize this framework.

The instances of fiction to which we will apply these questions—the "fi" in this book's fi-sci case studies—are not primarily about science phenomena, nor do they necessarily feature scientific discourse. In this way they categorically differ from the narratives typically discussed with relation to interfaces of literature and science. Often those are either science fiction, long employed as a vehicle for envisioning alternate worlds and futures, like Edwin Abbott Abbott's *Flatland* (1884), Octavia Butler's *Dawn* (1987), and Margaret Atwood's *Oryx and Crake* (2003), or fiction that does not present as science fiction but that still explicitly deals with or references science, like Jeanette Winterson's *Gut Symmetries* (1997) and Richard Powers's *Orfeo* (2014). In these works the connections between fiction and science are right there in plain sight. Fi-sci, on the other hand, takes an altogether different approach to the relationship between and intersection of fiction and science, and thus can be applied to instances of fictionality that bear no relation to science. In fact, the fi-sci model turns on latency. As I illustrate with examples at the conclusion of this chapter, rather than searching fictive discourse for literal manifestations of a given science phenomenon (e.g., as when the science phenomenon itself presents in the plot), we will look for figurative manifestations of this concept (the *pattern* of the science phenomenon is mirrored in some aspect of the fictive discourse). Fi-sci pattern mapping requires excavation.

•

With its capacity to illuminate even the most counterintuitive natural phenomena, the fi-sci framework can be applied whenever we are faced with science—in the classroom, in public outreach, and in any context that calls for science communication. So while drawing analogies between fiction and science may itself come across as counterintuitive, I aim to show that the method's potential to make science more accessible and enhance overall apperception supports this seemingly peculiar adventure.

The Fi-Sci Model

The following breakdown depicts the fi-sci pattern-mapping model's components and the relationships among them. It presents the fi-sci framework as the result of using the fictionality instrument in conjunction with the pattern-mapping cognitive maneuver—specifically, mapping from the source of fiction onto the target of science:

Instrument: Rhetoric is the art of using words (or other signs) as a means to an end—to motivate, to galvanize, to captivate. At times, we use words fictionally: We invent in order to—nondeceptively—get our audience to imagine things that are not real. For example, if your friend tells you they're going to take their starship to make your meeting on time, they are using fictionality as a means to the end of conveying that they will get there as quickly as possible—and perhaps channeling the humorous absurdity of the scenario into your forgiving their potential tardiness. In short, they are using fictionality rhetorically.

+

Cognitive maneuver: Identifying commonalities between two concepts entails recognizing that they share a pattern. Blood vessels, tree branches, and river deltas vary greatly in scale and function, but they have something in common: All three systems take an arborescent form. Pattern mapping is at play when we use the language of one discipline or domain to describe another. For example, we talk about recombinant DNA and genetic engineering with words that typically characterize linguistics and text, such as "code," "alphabet," "phrase," "editing," "instructions," "transcription," "translation."

=

Framework: Let's borrow language from the domain of fiction to elucidate concepts in the domain of science.

In this introductory chapter I weave between the model's components (indicated with brackets in section headings), elaborating on each, defining the terms involved, and analyzing their implications for this study.

Analogy: "The Fuel and Fire of Thinking"

Analogies and metaphors[3] do much more than provide literary decoration; "Metaphorical thought is a distinctive mode of achieving insight, not to be

3. Though "metaphor" and "analogy" are not synonymous, they will be used interchangeably for the purposes of this book. Suffice it to say that I regard the category of "metaphor" as a subset of the category "analogy": All metaphors are analogies but not all analogies are metaphors.

construed as an ornamental substitute for plain thought" (Black 237). When Romeo declares that "Juliet is the sun" (379), he (or Shakespeare) does more than adorn his description of Juliet with imagery. By positing a similarity—even an exaggerated identicality—between Juliet and the sun, Romeo (or Shakespeare) calls on the audience to apply some aspects of their knowledge of the latter to their mental construction of the former. The insight thereby achieved is a more nuanced understanding of what Juliet means to Romeo.

While this kind of reasoning may seem straightforward, "it is only relatively recently that metaphor has been reinterpreted as primarily a cognitive process" (Sullivan 1). I. A. Richards is recognized as the first to propose that we treat metaphor "as not only a linguistic matter" but as "part and parcel of human thinking" (Carrillo and Martínez 239). "Throughout the history of Rhetoric," writes Richards, "metaphor has been treated as a sort of happy extra trick with words. . . . a grace or ornament or *added* power of language, not its constitutive form" (90). Contrary to the "limited traditional view of metaphor" (93), Richards insists that it is "the omnipresent principle of language" (92). He contrasts the conventional conceptualization of metaphor with the one he now proposes as follows: "The traditional theory. . . . made metaphor seem to be a verbal matter, a shifting and displacement of words, whereas fundamentally it is a borrowing between and intercourse of *thoughts,* a transaction between contexts" (94). Returning to the above example, *shifting and displacing words* refers to using "sun" in place of "Juliet"—or at least the implication that audiences could do so without relinquishing accuracy. Recognizing this metaphor as a *transaction between contexts,* however, reveals how it leads audiences to map a pattern from the context of celestial bodies onto the context of objects of affection. Thus one domain illuminates the other.

Max Black further reconceptualized the roles and possibilities of analogy in our understanding by relating it to scientific models: "Those who see a model as a mere crutch are like those who consider metaphor a mere decoration or ornament. But there are powerful and irreplaceable uses of metaphor not adequately described by the old formula of 'saying one thing and meaning another.'" Specifically, a metaphor "has the power to bring two separate domains into cognitive and emotional relation by using language directly appropriate to the one as a lens for seeing the other" (236). This applies to the case studies explored in this book, which, though not scientific models in the conventional sense, function in a similar manner in that models are like avatars: Both provide instantiations of a core idea. In the models-avatars we will consider—the fi-sci analogies—language directly appropriate to one domain will serve as a lens for seeing the other domain, which can help us conceptualize the contours of what otherwise may be inconceivable. For example, viewing the ontological ambiguity of the Muppets as an avatar of quantum superposition gives us something that quantum superposition is *like*. Having seen something it is like, we can proceed to think in terms of transacting between the contexts of fiction and science.

In 1980 George Lakoff and Mark Johnson published the definitive study of metaphor in cognition, demonstrating that "our ordinary conceptual system . . . is fundamentally metaphorical in nature" (3). We automatically and ceaselessly categorize every concept we come across, and analogy is the mechanism that carries out this categorization. Faced with a novel concept, we reason about it by comparing it to something already familiar (Hofstadter and Sander 28). "*Thought* is metaphoric, writes Richards, and proceeds by comparison" (94). Hofstadter and Sander foreground the indispensability of analogy: "Without concepts there can be no thought, and without analogies there can be no concepts" (3); analogy is "the core of cognition"; "the fuel and fire of thinking" (3). Thinking *is* analogy-making.

Neil deGrasse Tyson posits that making analogies is what allowed humankind to endure. Describing protohumans, he remarks, "The best thing we had going for us was our intelligence, especially our gift for pattern recognition, sharpened over eons of evolution. The ones who were good at spotting prey and predator, telling poisonous plants from the nourishing ones, they had a better chance to live and reproduce" ("When Knowledge Conquered Fear"). Such spotting and telling are the products of comparison: "In order to survive, humans rely upon comparing what's happening to them *now* with what happened to them in the past" (Hofstadter and Sander 28). So we think by making analogies and survive by making analogies. In short, analogy is inseparable from cognition.

Target Domain: Science | Differentiating Perceptual Groups

Hofstadter and Sander maintain, "No thought can be formed that isn't informed by the past; or, more precisely, we think only thanks to analogies that link our present to our past" (20). What are we to do, then, when faced with a situation that we are unable to link to our past, because there is nothing in our past that is like this situation? "Any area for investigation," writes M. H. Abrams, "so long as it lacks prior concepts to give it structure and an express terminology with which it can be managed appears . . . inchoate—either a blank, or an elusive and tantalizing confusion" (31). Baffling science is precisely such an area of investigation: We cannot immediately tie it to prior concepts and thus lack terminology with which to reason about it. Hofstadter and Sander echo Abrams, who explains that when we try to understand something new, we look for something familiar to which we can compare it: We "cast about for objects which offer parallels to dimly sensed aspects of the new situation, to use the better known to elucidate the less known, to discuss the intangible in terms of the tangible" (31–32). In other words, we must find a comparable concept so as to make an analogy between the old and the new.

The problem is that casting about for objects that offer parallels to a new situation seems nigh impossible when said situation is counterintuitive. Must we therefore resign ourselves to never being able to form thoughts about such a situation? I believe we can escape that supposed fate by taking two steps. Step (I) involves making a small adjustment to our navigation: Let us cast about for objects that offer parallels a little further afield—in fact, in another field altogether. The analogies that grab our attention the most already cross concept/entity types—often, that's what makes them stand out: Juliet being compared to the sun is striking because a *person* is being compared to a *star*, two very different types of things. But as we have seen, not all analogies are figurative like that one; we are constantly making literal, perhaps less noticeable comparisons. For example, standing in front of a building you're seeing for the first time, you know that (a) the panel-like object built into its facade is a door, and (b) you should open it to gain entrance, because you are automatically comparing this panel to similar ones you've seen in similar circumstances. Recognizing that the object is a door because it resembles doors in your past is certainly not a cross-type analogy. So though not all analogies require that we look afield, soon I hope to show that in order to conceptualize counterintuitive concepts, we should do just that.

Step (II) is a cognitive move. It is something extra the mind can do to gain access to phenomena that are seemingly unlike anything else, phenomena that resist ties to a source domain and defy comparison to anything readily familiar. Before examining this something extra, however, let's acquire a better understanding of the parameters; let's explore the ocean of perception and locate where the more confounding objects reside.

The interface between the human mind and the natural phenomena that surround it—what David Herman calls "person-environment interactions" (ix)—can be parsed into two perceptual groups: phenomena that are intuitive, such as photosynthesis, evaporation, and Newton's Laws of Motion ([G1]); phenomena that are counterintuitive, such as eleven-dimensional spacetime, length contraction, and quantum entanglement ([G2]). While there are further gradations and types of perceptual complexity—soon we will split [G1] into subgroups [G1a] and [G1b]—this binary split makes the fundamental distinction between phenomena the human mind can readily conceptualize and phenomena the human mind struggles to conceptualize.

We begin to examine what sets apart complex and counterintuitive science from the easier-to-understand kind by exploring the distinctions among perceptual groups and subgroups. The behavior of systems in subgroup [G1a] is *straightforward*; they behave as we would expect them to. For example, playing billiards, one who has no training in physics can intuit the trajectory of a ball; math equations and physics formulas will unfailingly support this intuition. The behavior of systems in subgroup [G1b], on the other hand, is more *complex*. It is not beyond what we are capable of visualizing, but due to factors such as scale and scope—both spatial and temporal—these systems may pose more challenges to our understanding than ones that behave in clear, obvious ways. One such system is evolution by natural selection (EbNS); to see why it poses challenges to our understanding, consider the relationship between comprehensibility and narratability.

Herman observes that across wide-ranging disciplines beyond narratology, such as philosophy of history, psychoanalysis, psychology, and evolutionary theory, there are scholars who adopt some version of the notion that narrative is a sensemaking tool and that we understand the world and our experiences through story (4–5). An extreme version of this view is offered by Roger C. Schank and Robert P. Abelson, who claim that "all knowledge is encoded as stories" (2). H. Porter Abbott disagrees with such a universal formulation, but he believes "it is nonetheless easy to overlook just how pervasive storytelling is in our thinking and in the communication of our thoughts." Abbott suggests we have "a genetic predisposition to grasp events in time through story form" ("Narrative and Emergent Behavior" 229–30), and that "narrative is the principal way in which our species organizes its understanding of time" (*Introduction to Narrative* 3).

How does our proclivity to understand via storytelling affect our ability to imagine EbNS? First, Abbott observes that as an emergent behavior—described as "the coming into being of objects or patterns that . . . are the result of innumerable local interactions" ("Narrative and Emergent Behavior" 228)—the process of EbNS lacks three qualities that would have facilitated our understanding thereof: centralized control, clear causality, and, most importantly, a "narratable thread," which readers follow when they pick up on a "coherence . . . of events" (233). Wanting agential characters, a controlling, guiding entity, causative intention, and the crucial thread, EbNS has scant narratability. Abbott adds that "emergent behavior presents a special challenge to our need (or desire) to perceive narrative 'belonging,' that is, some sense of how the details of events in time make perceivable a story of change" (237). In short, the emergent nature of EbNS makes it difficult for us to experience this process as a story, which, in turn, makes it more challenging to imagine than a scientific process that is more straightforward.

Though understandable to different degrees, what phenomena in subgroups [G1a] and [G1b] have in common is that they come across as intuitive; they cohere with what we would expect given our everyday experiences. The behavior of systems in [G2] quickly becomes murky and is often counterintuitive. We describe interactions and behaviors of objects in this group using quantum and relativistic mechanics. Let's consider these in turn. In the quantum realm, matter exhibits wave-particle duality, whereby, writes Johnjoe McFadden, "photons may be emitted as particles, they may be detected as particles, but when unwatched, they travel through space as waves" (157); cause-and-effect seems meaningless, as "events at the level of fundamental particles do not have causes" (172); curiously, it is the measurement of a quantum system that seems to force the system to choose a state that makes sense to our everyday understanding. Unlike the trajectory of a billiard ball, the behavior of particles does not correspond to embodied experience and observed reality.

The behavior of objects at the cosmic scale also manifests in phenomena that are strange to human cognition. The theory of relativity accounts for some very odd features of spatial geometry and the passage of time: Two parallel lines can meet, and a line is not the shortest distance between two points; persons traveling at different speeds age differently. These are not outcomes one would intuitively expect. Admittedly, the above do not seem *as* bizarre as the behavior of quantum objects. We are able to conjure up two parallel lines meeting if we mentally curve the space we are picturing such that it forms a sphere—much like we see how longitudinal lines that appear to be parallel to each other on a flat atlas meet at the poles when we fashion the atlas into a globe. But our resorting to cognitive acrobatics to approach this phenomenon shows that it at least tends toward incomprehensibility. To our everyday senses, space is Euclidean, and therefore a phenomenon that requires us to imagine non-Euclidean geometry seems perplexing.

The ocean of perception is vast; the entities that make it up are not equally comprehensible. Now that we can better fathom its fathoms, let's return to my proposal that we take two particular steps when attempting to reason about phenomena that defy comparison to anything readily familiar. Step (I), I suggested, entails looking for parallel-offering objects in a field other than that of the target, which likely requires that we use figurative, rather than literal thinking. In what follows, we explore how scientists have translated ideas across disciplines—how they've used the language of one field to illuminate concepts in another.

[Cognitive Maneuver] "The Quest for the Invariant"

Since we cannot visualize the counterintuitive, we must approach accessibility to confounding science from a different angle. That a phenomenon is counterintuitive means that it is not similar enough to something else in our mental repertoire to be familiar or recognizable. We do not readily connect it to an existing concept and may feel stymied when we attempt to do so. Let's therefore see how we might generate a connection; how we might locate a site on which to build a bridge from one concept to another. The meta-analogy of the bridge is of course not new. Social anthropologist Claude Lévi-Strauss characterizes this matter of building a bridge of understanding as "a problem . . . of translation, of translating what is expressed in one language—or code . . . into expression in a different language" (9).[4] "Nature," he reflects, "has only a limited number of procedures at her disposal," so "the kinds of procedure which Nature uses at one level of reality are bound to reappear at different levels" (10). We can read different "levels" as different systems, scales, disciplines, and so forth.

Seven decades after Lévi-Strauss made his observation about patterns in nature, professor of mechanical engineering Adrian Bejan codified nature's "repeated procedures" into the constructal law of physics. The constructal law describes and predicts the direction of flow in an efficient system. For instance, consider river deltas, lung passageways, and lightning bolts (Bejan and Zane 1). The current flowing through each respective system is water, air, and electricity. But though these flow systems vary greatly in size and in many of their properties, they have one thing in common: currents flow through them in an arborescent pattern. Other systems' currents take other shapes: cracks in dry earth spread in the same pattern as a solidifying snowflake and a liquid droplet's splash (5); a long silk scarf dropped from a high balcony "undulates so that it falls like a meandering river" (13);[5] the "chimneys of steam escaping" a pot of boiling pasta or rice form the patterns of volcanoes, which likewise conduct heat and steam (13). Bejan and J. Peder Zane

---

4. Several pages later Lévi-Strauss clarifies, "I do not mean a different language like French or German, but different words on a different level. After all, this translation is what a dictionary is expected to give you—the meaning of the word in different words, which on a slightly different level are isomorphic to the word or expression you are trying to understand" (12).

5. The pattern of an undulating scarf pertains to a local segment of coursing water. The current flowing through river deltas is also water, but because there the focus is on the shape of the entire delta—afforded by a satellite view—the pattern is different: It is arborescent. Thus, even when the currents in question are the same, the parameters we assign a given system can affect what pattern emerges.

explain, "All flow systems generate shape and structure in time in order to facilitate this movement across a landscape filled with resistance (for example, friction). The designs we see in nature are not the result of chance. They arise naturally, spontaneously, because they enhance access to flow in time" (3). The constructal law shows why "'designedness' (configuration, rhythm, scaling rules) happen[s] *everywhere* in animate and inanimate systems alike" (14). Bejan hereby provides a scientific reason for patterns repeating across systems that vary greatly in current, function, and scale.

While Bejan studies evolution in physics, Lévi-Strauss offers the workings of the genetic code as an example of a natural procedure that "reappear[s] at different levels": "It is well known that, when the biologists and the geneticists had the problem of describing what they had discovered, they could do nothing better than borrow the language of linguistics and to speak of words, of phrase, of accent, of punctuation marks, and the like. . . . it is the same kind of problem arising at two different levels of reality" (10). What Lévi-Strauss terms "procedures" are essentially structural patterns; what he terms "different levels of reality" are different perspectives or disciplines; the translation he describes between linguistics and genetics is pattern mapping. When we cross the bridge of analogy, we "borrow the language" of one domain in order to expound a phenomenon in another domain. We map patterns in our "quest for the invariant" (8). And though Lévi-Strauss does not limit his discussion to science, his statement bears a marked resemblance to J. Bronowski's assertion that "all science is the search for unity in hidden likenesses" (13).

James Gleick's research in communication supports Bejan's and Lévi-Strauss's perspectives on pattern repetition. Gleick shows that as molecular biology was forming in the 1940s, scientists spoke of their discoveries using the language of information: "Codes, instructions, signals—all this language, redolent of machinery and engineering, pressed in on biologists like Norman French invading medieval English" (288–89). Gleick recounts that when Schrödinger described genetic information, he "summoned an example not from wave mechanics or theoretical physics but from telegraphy" (288):

> It has often been asked how this tiny speck of material, the nucleus of the fertilized egg, could contain an elaborate code-script involving all the future development of the organism. . . . For illustration, think of the Morse code. The two different signs of dot and dash in well-ordered groups of not more than four allow of thirty different specifications. Now, if you allowed yourself the use of a third sign . . . and used groups of not more than ten, you could form 88,572 different "letters." (Schrödinger 61)

And Schrödinger's summoning of such language is not a lone case. "Some of the physicists now turning to biology," continues Gleick, "saw information as exactly the concept needed to discuss and measure biological qualities for which tools had not been available: complexity and order, organization and specificity." (289). Neither tools, nor suitable words, it seems. In 1950 theoretical physicist Sidney Dancoff described a chromosomal fiber thus: "The entire thread constitutes a 'message.' This message can be broken down into subunits which may be called 'paragraphs,' 'words,' etc." (qtd. in Gleick 289). In all these examples, DNA is conceptualized as a linguistic text.

Strikingly, Gleick reflects that by the time James Watson and Francis Crick discovered that "the precise sequence of the bases is the code which carries the genetical information" (qtd. in Gleick 292), "they were no longer speaking figuratively" (292). We see that what started as an analogy evolved into the thing itself; the words that began as analogs, "translating what is expressed in one language . . . into expression in a different language" (Lévi-Strauss 9), in effect transposing meaning from a different domain, became genetics' native language. Gleick marvels, "By now the word *code* was so deeply embedded in the conversation that people seldom paused to notice how extraordinary it was to find such a thing—abstract symbols representing arbitrarily different abstract symbols—at work in chemistry, at the level of molecules" (295). Eventually, "the technical jargon of biologists included the words *alphabet, library, editing, proofreading, transcription, translation, nonsense, synonym,* and *redundancy.* Genetics and DNA had drawn the attention . . . of classical linguists" (299). In sum, to talk about genetics, one uses the language of information—language we use when we speak about text.

Physician Siddhartha Mukherjee substantiates this particular transposition of language in *The Gene* (2016). He asks that his readers "think of the genome as the *encyclopedia* of all genes, with *footnotes, annotations, instructions,* and *references*" (11, emphases added) and characterizes our drive to solve the genetic code as a "desire to decipher our manual of instructions" (12). Describing transcription, Mukherjee makes a lovely analogy:

> The process was akin to a library of rare books that is accessed for translation. The master copy of information—i.e., the gene—was stored permanently in a deep repository or vault. When a "translation request" was generated by a cell, a photocopy of the original was summoned from the vault of the nucleus. This facsimile of a gene (i.e., RNA) was used as a working source for translation into a protein. (166)

Using the language of one domain to describe something in another domain is exactly what happened in the case of the wave. Hofstadter and Sander write that around 240 BCE, the philosopher Chrysippus considered the possibility that sound is a sort of wave. Two centuries later the architect Vitruvius likened the way sound waves spread to the way ripples move on water. "What Vitruvius did," Hofstadter and Sander observe, "is in fact extremely typical of the thinking style of all physicists: taking a familiar, visible, everyday phenomenon and seeing it, in one's mind's eye, as taking place in another medium, sometimes at a vastly different spatial or temporal scale, so that it is inaccessible to one's senses" (210). Soon, scientists made the leap from sound to light, which meant that "all the standard old concepts associated with earlier waves could be investigated—*wavelength, period, speed, transverse* or *longitudinal, interference, reflection, refraction, diffraction,* and so on, and many of the same equations carried over beautifully from one medium to another" (213). The "standard old concepts associated with earlier waves" (i.e., the terms used in what we might call the language of waves) could be applied to (used to reason about and discuss) analogous concepts arising in a separate domain, "revealing a fruitful network of similarities between phenomena" (454).

Lévi-Strauss, Bejan, Gleick, Mukherjee, and Hofstadter and Sander, describing different eras and fields, all attest to the same technique: In order to conceptualize our world, we map patterns from one domain onto another. Lévi-Strauss says that "if we look at all the intellectual undertakings of mankind, as far as they have been recorded all over the world, the common denominator is always to introduce some kind of order" (12–13).[6] This inclination is echoed in Gestalt principles, a fundamental tenet of which is humans' "natural compulsion to find order in disorder" ("What Are the Gestalt Principles?"); to find it, we build "an analogical bridge between a current percept and a stored structure" (Hofstadter and Sander 455). In mapping patterns, we bring order to disorder; we show that things that seem arbitrary, random, or different actually share a form.

You may notice that none of the science examples in the present section—from electricity flow to water flow, from the genetic code to sound waves—is counterintuitive. Indeed, before delving into the more extreme cases of conceptualizing the seemingly incomprehensible, we observe that when constructing links of understanding from something already known to something new of a different type, we use the language of one to illuminate the other. We've seen that this kind of transfer turns on figurative thinking: connecting things that share properties but are not literally alike.

6. See also Campbell.

Source Domain: Fiction | Category Evolution

Having glimpsed how scientists use the language of a certain field to illuminate concepts in a different field, let's see if we might follow a similar approach in our search for phenomena that parallel counterintuitive science. But here we face a significant hurdle: One could argue that counterintuitive notions cannot, by definition, be linked to any structure that we would even recognize. Indeed, McFadden points out that "our brains—used to dealing with classical-level phenomena—are ill-equipped to form models of the quantum world" (175). When Werner Heisenberg says that analogies "may be justifiably used to describe things for which our language has no words" (10), he is talking about something beyond the struggle to express novelty. He is instead indicating neurological incapability. Arkady Plotnitsky explains such incapability with relation to non-Euclidean mathematics. Euclidean thinking "reflects the essential workings of our neurological machinery. . . . our *thinking in general,* as the product of this machinery, is Euclidean. . . . that which is beyond the uncircumventable limits of thought and knowledge . . . is beyond the capacity of our thought. It is literally unthinkable" (410). Heisenberg locates the problem in language and visualization: "It is not surprising that our language should be incapable of describing the processes occurring within the atoms, for . . . it was invented to describe the experiences of daily life. . . . words can only describe things of which we can form mental pictures, and this ability, too, is a result of daily experience" (11). It seems words do not exist in our neurological machinery to correctly describe incomprehensible phenomena.

If counterintuitive notions cannot be linked to recognizable structures, how will we ever find phenomena that parallel counterintuitive science? I believe the problem is not that corresponding structures do not exist. Rather, they reside in a realm so seemingly remote from science, we may not have considered scouting it for a spot to anchor the other end of the bridge. This realm is fictionality. Because some of the scientific concepts we are dealing with range from challenging to outright counterintuitive, in order to provide a fitting "home" for them, the source domain—the language whose words we will borrow—must be unbound from our daily experience. A language that bears these qualities is fiction.

To pinpoint exactly how mapping a pattern from fiction to science offers greater conceptual access to the latter, let's start by exploring how general conceptualization comes about. Hofstadter and Sander show that conceptualization is a process of extending our mental categories, and that this extension is powered by analogy-making. According to their model of category extension, conceptualization begins in infancy with single-member categories, for example, "Mom." Through a "long series of spontaneous analogies" (Hofstadter and Sander 39), this single-member category of "Mom" extends to include multiple members, like "Dad"; "Nanny"; "Teacher." Each of these is analogous to "Mom" in that they share prominent properties. Over time, the category "Mom" extends to encompass abstractions such as "Mother nature"; "Motherboard"; "Mother tongue"—concepts that also share properties with "Mom," but in a more figurative manner.

Hofstadter and Sander claim that "the very same mechanisms underlie" (39) the two cognitive acts: the analogies our minds create to get from single-member categories (let's call this [A]) to multiple-member categories ([B]), and the analogies our minds create to get from [B] to abstractions ([C]). I agree that analogy underlies the two cognitive acts but believe that because [C] requires figurative thinking there is a qualitative difference between [A]/[B] and [C]. We arrive at both "Sister" ([B]) and "Mother nature" ([C]) by analogically extending the properties of "Mom," but the extension to "Sister" does not require the degree of inventiveness and imagination that the extension to "Mother nature" does. Therefore, rather than placing both acts under the heading "category extension," I rename the move from [A] to [B] category *expansion* (a change in quantity), and the move from [B] to [C] category *evolution* (a change in type). Category evolution is the process by which figurative thinking offers greater conceptual access to phenomena—and often this figurative thinking leans on our capacity for fictionality. For instance, to perceive a circuit board or a language as a mother, we use a form of "what if" musing to imagine nonactual states. Unlike the case of [B], in [C] figurative thinking allows us to carry over certain properties of "Mom" even to some entities that are not living organisms.

Thus I augment and modify the Hofstadter-Sander model to account for an additional layer I identify in the move from [B] to [C]: figurative thinking that is often facilitated by the mind's capacity for fiction. The diagram in figure 1 depicts the progression from [A] to [B] to [C]. The augmented model supports the hypothesis that in order to increase comprehension and conceptualization, we may find it useful—and sometimes perhaps even necessary—to bring fictionality into play. Our capacity for invention helps us access what I've renamed figurative analogies, listed in category [C] (see figure 1).

This additional layer of figurative thinking that activates category evolution takes us partway toward step (II), the cognitive move we've been gearing up to define, the one we make to reason about phenomena that are seemingly unlike anything else. It turns out that even this going partway is already helpful to our reasoning about straightforward phenomena. Indeed, using

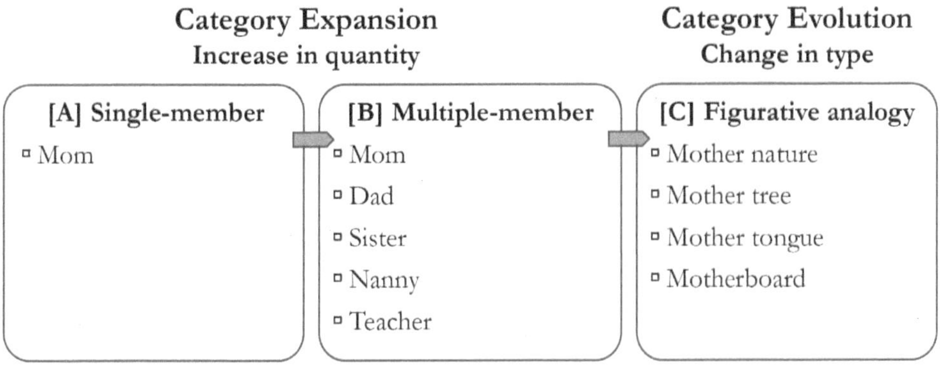

FIGURE 1. Conceptual category evolution.

figurative language is not limited to counterintuitive concepts. But in attempting to access counterintuitive concepts, we are helped by figurative thinking. With this, I rephrase my hypothesis so that it is more specific: When we use the language of one domain to illuminate a domain of another type, and that target domain is counterintuitive, it benefits us immensely when the source domain is fictionality.

What Kind of Science?

Now let's incorporate science in figure 1. To do this, we first split the figurative analogies listed in category [C] into an analogy's two constituents: the source and the target. Accordingly, figure 2 lists science phenomena below the target constituents of the analogies in category [C].[7] Moving from the top down, the perceptual groups in this column of targets are listed in order of increasing perplexity.

Figure 2 illustrates my proposal that our understanding of any kind of science—even that which is straightforward—can be enhanced with figurative analogies, with recourse to the instrument of fictionality. But rhetorical fictionality becomes ever more useful at higher levels of perplexity: It is valuable for better conceptualization of complex science, and perhaps key when it comes to our accessing counterintuitive science. Counterintuitive notions are unlinkable to the embodied experience of the world. Since fiction is unconstrained by correspondence to embodied experience, it makes for an excellent

7. Figure 2 replaces the previously used "G" (for "group") with "C," which denotes the figurative analogy component of conceptual category evolution. Thus, [C1a] corresponds to [G1a]; [C1b] corresponds to [G1b]; [C2] corresponds to [G2].

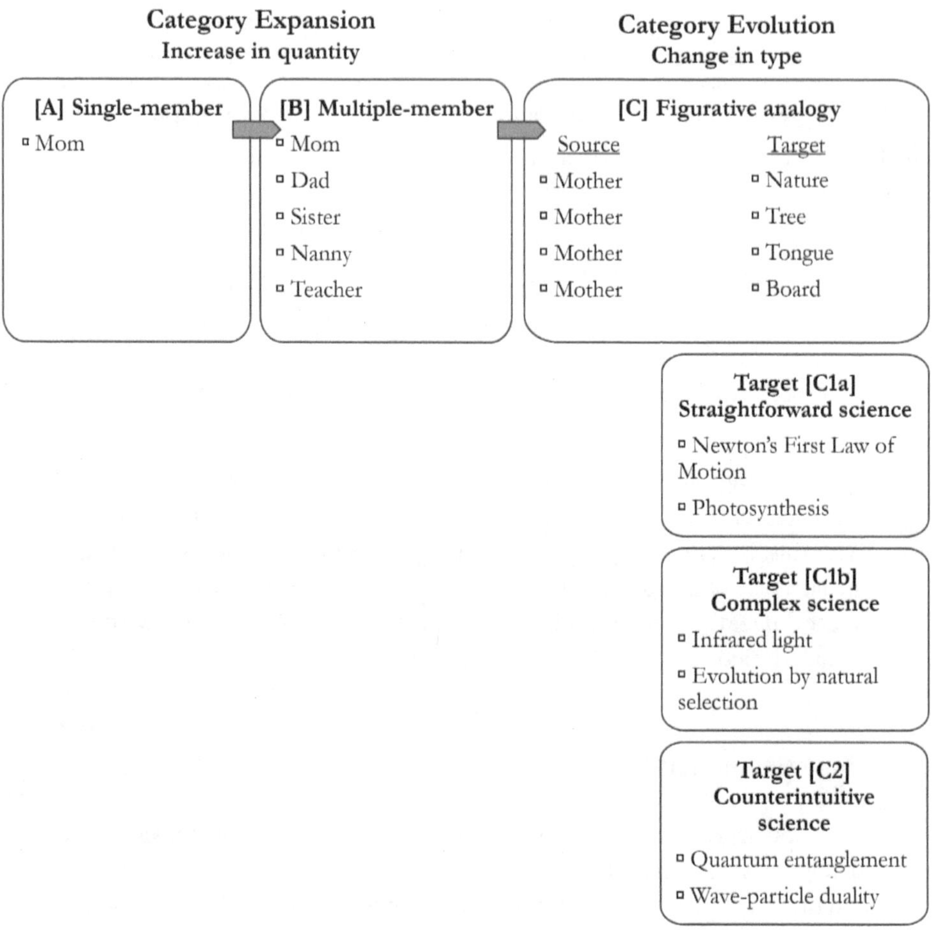

FIGURE 2. Conceptual category evolution plus science perceptual groups.

language for accessing the counterintuitive. Fi-sci pattern mapping demonstrates that access to confounding phenomena turns on the same mechanism that grants access to *all* concepts: analogy. The crucial difference is that when it comes to the counterintuitive concepts in [C2], a productive and sometimes pivotal source domain is fictionality. Let us approach confounding phenomena by drawing on our cognitive faculty for invention.

WHAT KIND OF FICTION?

This book's aim is to show how fiction illuminates science. So while science lives in the target domain of our study, fiction lives in the source domain. I reckon that upon analyzing fi-sci analogy case studies, we will find that the kind of fiction that facilitates our comprehension of strange scientific phenomena varies with the kind of science in question. But first: What kinds of fiction are there?

David Gorman defines fiction as "one kind of intendedly but non-deceptively untrue discourse" (163). I propose that untrue discourse can be untrue in different ways, two of which relate to our purposes: (a) It can be untrue in that it is fantastical—not (currently) possible, but easily imaginable; (b) it can be untrue in that it is counterintuitive—untrue in a manner that does not merely bend the "rules" of reality, but breaks them.

The etymology of "fantastic" shows why it is a fitting term for this first way a discourse can be untrue. Consider the bolded words in the following definitions:

- *Oxford English Dictionary*: "from Greek *phantastikos,* from *phantazein* '**make visible**,' *phantazesthai* '**have visions, imagine**'"
- *Online Etymology Dictionary*: "from Greek *phantastikos* '**able to imagine**,' from *phantazein* 'make visible' (middle voice *phantazesthai* '**picture to oneself**')"
- *The American Heritage Dictionary of the English Language*: "from Greek *phantastikos,* **able to create mental images**, from *phantazesthai,* to appear"

While "fantastical" is commonly used to describe the magical or unreal, its etymology makes clear that something fantastical is nevertheless something *conceivable*—one can summon a mental picture of it. A gryphon is a great example of a fantastical creature: Though it is not possible in nature, we can easily imagine it. One of the reasons we can do so is that a gryphon is composed of parts with which we are familiar: those belonging to a lion, and those belonging to an eagle. Another reason we can do so is that the recombination of these familiar parts does not violate intuition. All we have to do to conjure a gryphon is rearrange already-known components in a slightly different way but, crucially, in accordance with the intuition that matches our perceived reality. Similar reasoning can be extended from fantastical creatures to fantastical events. Teleportation is not (yet?) possible, but it is imaginable because it combines two things with which we are already familiar in an intuitively acceptable way. Merging the concept of travel and the concept of instantaneous communication at a distance (as with a cell phone), we can imagine instantaneous travel at a distance. Again, it is a remixing of already-familiar concepts, which requires us only to bend possibilities beyond their present parameters.

On the other hand, an untrue discourse that is counterintuitive is a discourse that is anomalous. Consider the bolded words in the following definitions for "anomalous":

- *Collins English Dictionary*: "**deviating from the normal** or usual order, type, etc.; irregular, abnormal, or incongruous . . . from Greek *anōmalos* **uneven**, inconsistent, from an- + *homalos* even, from *homos* one and the same"
- *Kernerman Webster's College Dictionary*: "**not fitting into a** common, **familiar**, or expected type or **pattern**; unusual."

It is the deviation from a familiar pattern that I would like to focus on. Here, "pattern" is the shape of intuitive ontological understanding. Ontology, which is concerned with the nature of being and existence, is a branch of metaphysics explored by philosophers who seek to understand reality. The anomalous/counterintuitive in fiction is something that defies ontological rules.

At this point I would like to make a distinction between *counterintuitive* fiction and *unnatural* fiction. Though at first glance the two may appear identical, they describe disparate types of narrative. In showing how the two differ, I hope to clarify my view of the qualities that render some fiction counterintuitive. The usage of "unnatural" as a descriptor of certain narratives has a two-decade history. The impetus to study unnatural narratology stemmed from Monika Fludernik's 1996 monograph, *Towards a "Natural" Narratology*; as Abbott notes, "It would seem inevitable . . . that the possibility of an '*unnatural*' narratology would arise, especially given how thoroughly modernist and postmodernist writers have twisted, punctured, torn apart, and reversed what readers and viewers normally expect in narrative" (*Introduction to Narrative* 39). While there are differences in how unnatural narratology theorists define the unnatural, Alber et al. "stress a commonality: namely, an emphasis on the various ways in which certain narratives deviate from real-world frames by being highly implausible, impossible, unreal, or insistently fictional" (104). Alber in particular "restrict[s] the use of this term to texts about storyworlds that contain physically, logically, or humanly impossible scenarios or events" (449). Alber et al. emphasize, "The criterion we use to identify unnaturalness is actualizability, which bears on the question of whether the represented scenarios or events *could exist* in the real world or not" (104–5). While this may seem to conflate with my view of the counterintuitive in fiction, my definition bears a distinguishing qualification: To be characterized as counterintuitive, an element of fiction must not only be nonactualizable but also incompatible

TABLE 1. Islands in fiction
Terminology per Alber et al.

Counterfactual	Unnatural/Unactualizable	Unnatural/Unactualizable
"The fictional island in Daniel Defoe's *Robinson Crusoe* (1719), for example, could exist in the real world . . .	. . . while the flying island of Laputa in Jonathan Swift's *Gulliver's Travels* (1726/1735) could not" (105).	The island depicted in the television series *Lost*

Terminology per Trauvitch

Counterfactual	Fantastical/Unactualizable	Counterintuitive
Crusoe's island	Laputa	The island in *Lost*

with our embodied experience, observed reality, and intuition. Otherwise put, the counterintuitive is incompatible with embodied experience, but not every nonactualizable element is thus counterintuitive. A speaking cat is impossible (nonactualizable and unnatural) but much more easily imagined than a certain physicist's cat who exists in a state of quantum superposition (counterintuitive). To present the distinctions that follow from my qualification, I supplement the examples of islands used by Alber et al. (see table 1).

One departure in my interpretation is that I see the unactualizable and the fantastical as one and the same. Additionally, my qualification lets me distinguish between Laputa and the island in *Lost,* whereas the interpretation by Alber et al. would place both islands in the same group. In my view there are categorical differences among the three islands: The first is counterfactual, the second is fantastical (not possible), and only the third is counterintuitive—incompatible with everyday, observed reality. Recounting several bizarre instances in the opening episodes of *Lost,* Elizabeth Berkebile McManus concludes that "the castaways, and the viewer, are confronted with a space that does not adhere to the laws of nature." She cites the airplane's crash site a thousand miles off its course as an example that "the entire space lies outside the bounds of scientific knowledge and measurement" (7).

I do not consider Laputa counterintuitive because the thing that makes the island fantastical—its flight—is not incompatible with our intuition. Laputa is fantastical but easily conjured because, just like our gryphon, it is a novel combination of parts with which we are already familiar: (a) an island and (b) a flying entity. *Lost*'s island is incompatible with our embodied experience because—at least in one prevalent reading of the series—it is implied that this island exists in its own spacetime. For example, the show's depiction of

TABLE 2. Kinds of fiction

Type of Fiction	Intuitive—Possible	Intuitive—Fantastical, impossible	Counterintuitive
Example	Crusoe's island	Laputa	The island in *Lost*
Required of Reader	Imagine possibilities	Imagine impossibilities	Override intuition

on-island space vis-à-vis off-island space (in flashbacks)—part of what McManus calls "*Lost*'s unique embedded structure of space and narrative"—"seems to defy causality" (11).

Having distinguished between different kinds of fiction, let's distinguish what readers are asked to do in order to conceive of each kind of fiction. To conceptualize possible, realistic fiction, readers keep in mind that the story has not happened, but it could have. The premise of realistic fiction is that it represents what could possibly happen. All that is asked of readers is to be able to entertain such possibilities. To conceptualize fantastical, impossible fiction, readers are asked to do a little more: use their imaginations to envisage something that is not only counterfactual but also cannot actually happen. When it comes to counterintuitive fiction, however, readers must do even more—something quite radical. To conceive of a story that contravenes our embodied experience of the world, we need to *override an intuitive objection.* I use the term *override* here with three definitions thereof in mind:

- "To counteract the normal operation of (an automatic control)" (*American Heritage Dictionary*)
- "To take manual control of (a system that is usually under automatic control)" (*Collins English Dictionary*)
- "To modify or suspend the ordinary functioning of" (*Kernerman Webster's College Dictionary*)

In order to conceptualize an island like that in *Lost,* we must override what we know based on our observed reality, in this case, our embodied notions of spacetime and causality. These notions form the "system" that is "under automatic control." We have to "take manual control of"—"modify or suspend the ordinary functioning of"—what we intuitively know about spacetime and causality to conceive of the island on *Lost.* Table 2 depicts what's required of us as we confront each island.

The layer of figurative thinking I highlighted in the move from category expansion to category evolution took us partway toward the second step in my two-step proposal, the something extra the mind can do to gain access to counterintuitive concepts. We are now in a position to spell out the rationale for the second step: Step (II) is using fictionality as an instrument, since within the realm of fictionality we are capable of overriding our intuitive objections. At the start of this section, I hypothesized that the kind of fiction that facilitates our comprehension of strange science phenomena would vary with the kind of science in question. By the time we reach this book's coda, having investigated how seven examples of fiction map onto six examples of science, we will be in a position to say more about whether—and to what extent—the type of one tends to vary with the type of the other.

[Instrument] The Rhetoric of Fictionality

While storying science can help us understand it, in *fiction*—specifically fantastical and counterintuitive fiction that requires us to imagine impossibilities and override intuition, respectively—we can find avatars of complex and counterintuitive scientific phenomena. These phenomena require a flexibility of expression that the mode of fiction affords, flexibility that is enabled by the set of assumptions we bring to our reading of a fictional narrative. As Nielsen et al. observe, "The assumption that a text or passage is intended fictively radically alters our reception of it. . . . The assumption of fictionality . . . changes our interpretive activity and its outcomes" (67, 68). Indeed, there is a quality particular to fiction that can transport us to the confounding.

Narratologists examine fictionality from diverse angles, including cognitive, psychological, and aesthetic views, as well as from the one offered by possible worlds theory. The one most suited to the present endeavor is the rhetorical approach because analyzing fictionality as a "rhetorical resource" (Walsh 1) reveals its capacity to enable conceptual category evolution. The reason we are able to imagine the fantastical and the counterintuitive in fiction is that we are approaching it *as fiction*: We free our minds of the assumptions we would make when confronted with a nonfiction narrative—let's call these "nonfictive assumptions"—and replace them with "fictive assumptions" about the possible existence of entities and events that we tacitly understand are nonactual. In adopting fictive assumptions, we are able to approach the shape of the fiction's machinations—that is, accept the fantastical and counterintuitive things it presents. This acceptance opens our conceptual pathways to recognizing the same shape (what is now a pattern) in science.

One important feature of the rhetorical approach is that it distinguishes between fictionality as a general phenomenon and generic fiction, such as the short story, the novel, or the fiction film. In other words, the rhetorical approach views fictionality as a macro-genre of discourse and generic fiction as a subset of that macro-genre. This distinction underlies a related one that is important in my analysis: that between a fictional concept or entity, *x*—one that is independent of a narrative—and a fictional narrative, *y*, which entails change over time, in which *x* sometimes lives. Since *x* is independent of *y*, it can float free of story. The fi-sci framework draws on both *x* and *y*. For example, in my earlier discussion of the Muppets, I mentioned particular *y*s—such as *Muppets from Space*—but I also talked about Muppet*ness*: the quality of a Muppet, an entity that can present as fictional and as nonfictional, depending on the context. In the fi-sci analogy illuminated by the Muppets, which we explore further in chapter 5, the pattern mapping depends not on any one story they're in but rather on their conceptual status as ontologically ambiguous entities.

Let's return to our stalwart gryphon to illustrate the distinction between fictionality in generic fiction and fictionality as a general phenomenon. Ursula K. Le Guin's "Darkness Box" is a narrative, *y*, that features a gryphon, *x*. But there are also instances of *x* that exist independently of *y*. Like Athena issued fully grown and dressed from Zeus's forehead, the concept of a gryphon requires no (story-bound) gestation period. This legendary hybrid creature is evidence that fictionality need not begin in or be tethered to narrative. In exploring fi-sci, sometimes we'll rely on the fictionality of the story, and sometimes we'll rely on the fictionality of an entity or concept; either way we'll rely on fictionality. For it is not just the story that's available for our pattern-mapping needs; an entity or concept independent of or extracted from the story can contribute a shape in its own right. What the distinction between *x* and *y* does for the rhetorical approach to fictionality is take a clear position in the debate about whether fictionality entails narrative.[8] From the perspective of fi-sci, not only is fictionality separate from generic fiction, but it is also independent of narrative.

•

A given pattern's science avatar may be strange or counterintuitive, but because fiction by its very nature is free of reality's "limitations," and because it has provided a gargantuan repertoire of well-known inventions, it can give the pattern an avatar that *is* familiar, or that we *do* recognize. This freedom from limitations is instrumental to the move from [B] to [C] in figure 1. Without the affordances of fiction, our minds do not extend to the figurative thinking that enables [C]. To understand why, let's reflect on the matter

8. For a discussion of this debate, see the introduction in Gammelgaard et al.

of invention, which is central to the rhetorical definitions of fictionality proposed by Phelan ("intentionally communicated invention, projection, or other means of directing an audience to consider nonactual states or events") and by Henrik Zetterberg-Nielsen and Simona Zetterberg-Nielsen: "intentionally signaled, communicated invention" (Gammelgaard et al. 7). We can see that the quality of invention is not applicable to [B] but certainly applicable to [C]. When we characterize nature or a computer's main circuit board as a mother, we are intentionally signaling an invention and directing our audience to consider nonactual states.

Thus, fiction reveals truths even though it is "intendedly but non-deceptively untrue" (Gorman 163). "Where nonfiction involves a direct claim about its relevance to actual states of affairs . . . fictionality gets its purchase on the actual by inviting us to set aside the constraints of direct reference to the actual and assume invention or some other mode of indirection" (Gammelgaard et al. 5). Setting aside the constraints of direct reference to the actual is what makes it possible to express a counterintuitive concept's pattern as a relatable avatar.

[Framework] Expressions → Linguistic Schema → Flowchart

Having established how recourse to fictionality offers greater conceptual access to confounding scientific phenomena, let's take a closer look at the steps of fi-sci pattern mapping, that is, generating connections between phenomena in fiction to phenomena in science. I like to begin by identifying the pattern of the phenomenon in the target domain—science.[9] In the terminology of figure 1, for a given [C]—scientific phenomenon—we work backward to identify its [A]—the single-member category. This single-member category essentially conveys the concept's pattern. For example, we can distill the quantum superposition phenomenon into a pattern by using a linguistic schema consisting of three expressions. We start by establishing an expression that represents the science phenomenon's pattern, for instance,

A [quantum particle] is simultaneously {in state p} and not {in state p}.

This science (S) expression includes details specific to physics. I have placed the science terms in brackets because these are the terms we will attempt to

9. One can begin at a different part of the flowchart and go through it in any order; my students often do.

substitute with their analogs in another domain. I use different kinds of brackets—parentheses, square brackets, braces, angle brackets—around different terms. The particular kind of bracket I choose in each case is arbitrary; the brackets' purpose is to facilitate the conversion of one expression to another by visually representing like for like. So, the terms inside braces in the science expression should be analogous to the terms inside braces in the fiction expression; the terms inside parentheses in the science expression should be analogous to the terms inside parentheses in the fiction expression. The different brackets distinguish among sets of analogs. As I soon demonstrate, the different kinds of brackets indicate local analogies that when taken together form a global analogy.

We refine the S expression into a generalized version that represents just the phenomenon's shape, without the particulars of science. The template, or T expression, is a distillation of the S expression, the idea being that by working with a neutral iteration of the pattern we may more readily recognize its manifestation in fiction. To arrive at this T expression, we replace the science terms in the brackets with ones that are discipline-neutral, but we do not alter the words outside the brackets so as to preserve the relationships among the pattern's components:

An [entity] is simultaneously {p} and not {p}.

This neutral expression facilitates the substitution of terms in the target domain (science) with terms in the source domain (fiction). Such substitution renders the fiction (F) expression,

A [character] is simultaneously {fictional} and not {fictional}.

The three-expression procedure just showcased is a linguistic (as opposed to graphic) approach to pattern mapping; it is through manipulating linguistic expressions that we chart the pattern between the two domains, that is, locate the pattern's different avatars. We can attempt to course through the procedure and carry out this pattern mapping each time we encounter a science phenomenon. Though we cannot know in advance whether an F expression will materialize for every S expression—whether we will find a shape in fiction for every shape in science—when we do find an F expression we'll be able to wield it toward conceptualizing the science. The linguistic schema shows us where we need to go so as to amplify comprehension, and a direction we could take to get there.

The Pattern of This Book

Let's craft a graphic home for the three-expression procedure: Table 3 functions as a flowchart that incorporates a couple of pre- and postexpression steps, which get us from the science phenomenon all the way to the fiction phenomenon. A flowchart gives us a more detailed roadmap for the journey, a roadmap that doubles as the outline of this book: For each science phenomenon—each case study—our mission is to populate the pattern-mapping flowchart by following a seven-step process, which mirrors the organization of each chapter:

1. Identify a scientific phenomenon;
2. Identify an example of this phenomenon;
3. Establish an expression that represents this phenomenon's pattern, bracketing science terms;
4. Convert the science expression into a template expression by replacing bracketed science terms with discipline-neutral terms;
5. Convert the template expression into a fiction expression by replacing bracketed discipline-neutral terms with literary terms;
6. Identify a text of literary fiction in which the fiction expression manifests;
7. Identify the literary phenomenon exemplified in box 6.

As we complete these steps it is important to reiterate that the words outside the brackets ought not be altered:[10] The three expressions—boxes 3, 4, and 5—must be identical except for the bracketed words. The words outside the brackets relay the ways in which the bracketed words relate to each other: That is, they convey the structure of the phenomenon; that they're identical across boxes 3, 4, and 5 shows that the structure of the concept is the same regardless of domain.

TABLE 3. Pattern-mapping flowchart

	Phenomenon	Example	Expression
Science	1 →	2 →	3 ↓
Template	—	—	4 ↓
Fiction	7 ←	6 ←	5 ←

10. Exceptions are made for phonetic adjustments and grammatical correctness, e.g., adding/removing letters to maintain subject-determiner agreement.

In part 1 we will consider complex phenomena in biology. We'll locate the avatar of each phenomenon in fiction and demonstrate how such pattern mapping illuminates complex science. In part 2 we will encounter counterintuitive phenomena in nonclassical physics. Locating the avatar of each phenomenon in fiction, we'll explore how this pattern mapping opens access to counterintuitive science. Having established a fi-sci analogy, I will present it in the formulation

[Fiction Concept] || [Science Concept]

I use the double bar (||) with a nod to mathematics, where it indicates parallelism; the above formulation conveys that the fiction concept parallels—is analogous to—the science concept.

Tables 4 and 5 fill in steps 1–7 for each of this book's case studies, providing the results at the outset. These tables reveal the whodunit and dispel any mystery about conclusions, but in doing so they provide you, the reader, with a precise roadmap for the journey ahead, thus giving us more space to explore the results' implications. Moreover, glimpsing an aerial view of the destination upfront shifts the adventure's focus to the more intriguing question of *how* we'll get there.

By applying this kind of uniform organization to each chapter, I hope to gradually construct a solid case that fictionality can play a powerful part in our understanding of science; each completed fi-sci pattern-mapping flowchart strengthens this hypothesis. As for the progression of the chapters themselves, I begin with complex science and move on to counterintuitive science to show that the model works even as we challenge our minds to approach ever more perplexing phenomena. Overall, seeing how this kind of analysis can be applied across a range of science and fiction substantiates the model's versatility and scope.

Metaphorical Entailment

I have suggested that though we can certainly access intuitive natural phenomena without fiction, fiction both makes access easier and enhances our comprehension of the science. One of the mechanisms responsible for this enhancement is metaphorical entailment, which linguist Zoltán Kövecses describes as "rich additional knowledge about a source . . . mapped onto a target" (122). That is, not only can the source itself be mapped onto a target, but other properties of the source can also be mapped onto corresponding

TABLE 4. Aerial view of biology fi-sci case studies

	Chapter 1: Permutation	Chapter 2: Proliferation (Gametic)	Chapter 3: Proliferation (Clonal)
Science Phenomenon	Genetic engineering	Gametic reproduction	Clonal reproduction
Science Example	Editing the sequence of an organism's nucleobase pairs in DNA, thereby encoding different traits.	When dolphins reproduce, the offspring receives half of its chromosomes from each progenitor.	When aphids reproduce, each offspring sets out as a genetic replica of the progenitor and is identical to its siblings.
Science Expression	An [organism's genetic material] is recombined in order to generate variations in this {organism}.	In [gametic reproduction] multiplicity lies in {progenitors}, whose <DNA> converges in a new (organism).	In [clonal reproduction] multiplicity lies in iterations, each of which is a {copy} of a singular <progenitor>.
Template Expression	An [entity's constituent parts] are recombined in order to generate variations in this {entity}.	In [a certain type of proliferation] multiplicity lies in {predecessors}, whose <characteristics> converge in a new (entity).	In [a certain type of proliferation] multiplicity lies in iterations, each of which is a {version} of a singular <entity>.
Fiction Expression	A [story's segments] are recombined in order to generate variations in this {story}.	In [transtextuality] multiplicity lies in {precursor texts}, whose <voices> converge in a new (text).	In [authorial motifs] multiplicity lies in iterations, each of which is a {variation} on a singular <theme>.
Fiction Example/s	*Rayuela,* Julio Cortázar; *Malas hierbas,* Pedro Cabiya	*Adios, Happy Homeland!,* Ana Menéndez	Works by Jorge Luis Borges
Fiction Phenomenon	Resequencing story segments	Transtextuality	Authorial motifs

properties of the target. For example, earlier we saw that once scientists made the leap from wave as sound (source domain) to wave as light (target domain), properties of sound—such as interference, refraction, and diffraction—could be investigated as possible properties of light. Indeed, it turns out that "many of the same equations carried over beautifully from one medium to another" (Hofstadter and Sander 213), such that the property of refraction in sound maps onto the property of refraction in light. In these ways, the "elaborate knowledge about" the source "structures much of our knowledge about" the target (Kövecses 129). Each additional bit of information from the source that is *mappable* onto the target adds to an understanding of the latter.

TABLE 5. Aerial view of nonclassical physics fi-sci case studies

	Chapter 4: Warp	Chapter 5: Ambiguity	Chapter 6: Recursion
Science Phenomenon	Spacetime curvature	Quantum superposition	Environmental entanglement
Science Example	A wormhole can connect causally disconnected regions.	An electron is simultaneously in two different states.	A quantum system interacts with a drop of water.
Science Expression	[Spacetime] curves such that {regions} on either side of an effectively uncrossable boundary can become connected.	A [quantum particle] is simultaneously {in state p} and not {in state p}.	[Environmental entanglement] disrupts the {phase coherence} of a <quantum system> such that the <quantum system> cannot maintain its (quantumness).
Template Expression	[A framework] curves such that {entities} on either side of an effectively uncrossable boundary can become connected.	An [entity] is simultaneously {p} and not {p}.	[A mechanism] disrupts the {preservative} of an <entity> such that the <entity> cannot maintain its (unique properties).
Fiction Expression	[The storyfabric] curves such that {characters} on either side of an effectively uncrossable boundary can become connected.	A [character] is simultaneously {fictional} and not {fictional}.	[Mise en abyme] disrupts the {structural and ontological boundaries} of a <story> such that the <story> cannot maintain its (storyness).
Fiction Example/s	"Continuidad de los parques," Julio Cortázar	Jim Henson's Muppets	*Cien años de soledad,* Gabriel García Márquez
Fiction Phenomenon	Worm's-eye-view metalepsis	Ontological ambiguity	Mise en abyme

Note the emphasis on "mappable." Hofstadter and Sander's and Kövecses's qualifiers "many" and "much" are important: Not all properties of the source will necessarily map onto the target. The invariance principle pertains to those properties that do not map over: It "blocks the mapping of knowledge that is not coherent with the schematic or skeletal structure of the target concept" (Kövecses 131). Consider the motherboard. The video "What 'Motherboard' Really Means—And Why It Matters" explains that "just as a mother in a family nurtures, organizes, and connects the members together, the motherboard in a computer connects all the components and allows them to work in harmony" (01:53–02:05). Thus, one property of "mother" that maps onto "motherboard" is that of central connector. A property of "mother" that does *not* map onto "motherboard" is that of originator: Other circuit boards (aptly named "daughterboards") do not *come from* the motherboard; rather, they are components that attach to it. Kövecses might say about this originator "feature of the source" that it would be "prevented from being mapped onto the target" (132). In this way, the mind makes use of what works and does not apply what doesn't work.

Hofstadter and Sander equate category extension with analogy-making and maintain that "it is not merely for idle fun that one assign[s] [concepts] to a preexisting category in one's memory; it is principally because doing so gives one access to a great deal of extra information" (21). Metaphorical entailments—those properties that *do* map over—provide us with that much more knowledge or material that we can apply to the target domain. In a similar vein, Tasneem Zehra Husain remarks, "Not only are metaphors connective, they are also generative." She offers an example of "extra information" gleaned from Einstein's fabric metaphor for spacetime: "The mere comparison . . . suggests questions that were previously inconceivable: Can space-time be torn? Can it be punctured? Can it gather in folds? While gravity remained Newtonian . . . such lines of thought were not possible. They arise only in this new liminality" (23).[11] Thus, metaphorical entailments not only provide more information; they also *create* knowledge that would not have been reachable in their absence. Because metaphorical entailments offer such rich cognitive benefits, each of our case studies will feature a section about them as they pertain to the given fi-sci analogy.

Form

As we proceed in this investigation of what fictionality can do for our understanding of science, I propose that we approach and frame our analysis as one of form. Doing this facilitates our application of findings from existing research on formalism and new formalism, even as we add this book's particular angle to the larger conversation.

Earlier, I synthesized research by Lévi-Strauss, Bejan, Gleick, Mukherjee, and Hofstadter and Sander all pointing to this: Humans engage in pattern mapping so as to bring order to disorder—to show that things that seem arbitrary, random, or different, actually share a *form*. In *Forms: Whole, Rhythm, Hierarchy, Network,* Caroline Levine compares historical uses of the term *form* and notes a definition that these uses have in common: "'Form' always

11. This resonates with Black's view on the benefits of scientific models.

indicates *an arrangement of elements—an ordering, patterning or shaping*" (3). Forms are important "because they shape what it is possible to think, say, and do in a given context" (5). While Levine focuses on forms in the arenas of literature and politics, I posit that forms also shape what it is possible to think about *science*. Specifically, it is the recognition of the same form at play in both science and fiction that gives us access to new cognitive avenues.

Levine borrows from design theory the concept of *affordance*—which describes "the potential uses or actions latent in materials and designs"—and explores "both the particular constraints and possibilities that different forms afford" (6). Thinking about a form in terms of its affordances—the kinds of shapes that it allows—is precisely what we do when searching for an avatar of the science phenomenon in fiction. That is, our charge when it comes to each concept is to find a form in fiction that has a very particular potential latent in its design: the potential to illuminate the same pattern across disciplinary domains. Janine Rogers also explores this subject in *Unified Fields: Science and Literary Form*. To see "how form can be used to express scientific concepts" (xvi), she links poetic forms, like the sonnet and sestina, to scientific forms—for example, the forms of "the Pythagorean Theorem, the DNA double helix, and atomic structure." This allows Rogers to "demonstrate the essential *function* of form that is shared by both science and literature" (xvii). I maintain that this sharing of function has rhetorical consequences.

With these insights about form, let's return to a promise I made earlier: to show—with examples—why, rather than looking for literal manifestations of a given concept in fictive discourse, we will seek figurative manifestations. A literal manifestation of a science phenomenon is a matter of identicality, not analogy. For instance, if the science phenomenon is spacetime curvature, connecting it to an example in fiction where a character is experiencing spacetime curvature (and the ensuing time dilation, like Matthew McConaughey's character, Cooper, in *Interstellar*) is not making an analogy; it is saying, "Here, literally, is a manifestation of the science phenomenon in a work of fiction."

Instead, when searching for parallels in fiction, we ask, Where in fiction is there a form that shares contours with spacetime curvature? One answer to this particular question is the subject of chapter 4: warp. Both a wormhole and the metalepsis in Julio Cortázar's story "Continuidad de los parques" ["Continuity of Parks"] connect two entities that effectively should not be able to meet. It is not that a character in "Continuidad de los parques" enters a wormhole; the story has nothing to do with literal spacetime curvature. Rather, a character therein experiences a process that is *like* what happens when spacetime curves: The character intersects with something (another character) that intuitively should be in a different reality—in other words, out of reach.

This example illustrates the distinction: Our search is not for a literal manifestation of the science phenomenon but rather a figurative manifestation thereof. This figurative manifestation can arise in a narrative's structure or plot; in the reading or viewing experience; in the story's setup; via literary devices. The crucial bit is that the manifestation must not be of the idea itself, but rather of the idea's form.

Here's another example, this time related to chapter 1:

- Literal: Atwood's *Oryx and Crake* features genetically engineered creatures.
- Figurative: In Pedro Cabiya's *Malas hierbas* [*Wicked Weeds*], chapters are the figurative DNA of the story such that alterations to their sequence result in a different reading experience.

And an example related to chapter 3:

- Literal: The film *Oblivion* features clones.
- Figurative: Jorge Luis Borges's motifs are figuratively cloned throughout his body of work because they pop up in some configuration or another in a vast majority of his short stories.

The "fi" of the fi-sci analogies in chapters 1, 3, and 4 are the figurative ones presented above. Likewise, the "fi" in chapters 2, 5, and 6 are manifestations of forms that share contours with gametic reproduction, ontological ambiguity, and recursion, respectively. To round off the excavation analogy broached earlier, I add the following observation. In the context of fi-sci analogies, the literal manifestation of a concept shows up on fictionality's ground. The figurative manifestation of a concept—the manifestation of a concept's form—is below ground and must be excavated with pattern mapping. It is this unearthed manifestation that functions as the source that illuminates the target.

PART 1

BIOLOGY

CHAPTER 1

Permutation

We begin at the beginning with the constituent part of an organism: the gene, which carries inherited traits from one generation to the next. The gene is a segment of DNA, which is a molecule made up of nucleotides comprising (a) deoxyribose, (b) one phosphate group, and (c) a nucleobase. There are four nucleobases—adenine (A), thymine (T), cytosine (C), and guanine (G)—and because A always pairs with T and C always pairs with G, there are only four possible manifestations of nucleobases along a strand of DNA:

A	T	C	G
T	A	G	C

A gene is defined by the sequence of these four manifestations. It is the difference in the sequence of the four manifestations that leads to genetic variation. Gleick tells us that the matter is "not just information storage but information transfer. DNA serves two different functions. First, it preserves information. . . . Second, however, DNA also sends that information outward for use in the making of the organism" (296). This information is conveyed in the sequencing of the nucleobase pairs. Indeed, order is key: "You cannot read a book by dissolving all its words into alphabets," writes Mukherjee. "With DNA, as with words, the *sequence* carries the meaning" (216). Using recombinant DNA technology—also known as genetic engineering—scientists can alter the sequence of nucleobase pairs, and thereby convey *different* information, modifying the genome bearer's traits.

We cannot discern recombinant DNA, or rDNA technology, with the naked eye because it operates at a tiny scale. Those of its outcomes that *are* observable at the human scale take a considerable amount of time to become apparent, especially when it comes to organisms with a long lifecycle. Most significantly, these outcomes are not straightforward but complex: Manipulating the genetic makeup of cells in one part of an organism can result in unexpected changes in other parts. Genetic engineering is therefore less immediately discernible than a process that operates at a human scale and displays direct consequences. While comprehensible, rDNA technology is at any rate less accessible to the mind's eye than processes that we can see without external tools and whose effects are immediate, such as a tree losing its leaves. For a more comprehensive understanding of genetic engineering, let us turn to the imagination. Using the fi-sci framework, we will find that we can conceptually concretize the phenomenon of rDNA by mapping its pattern onto (that is, recognizing its avatar in) fiction.

Genetic Engineering

Recombinant DNA is "a type of DNA sequence that is composed of sequences from two or more different sources or organisms, such as human and pig sources, or synthetic (lab-made) sequences and microorganisms" ("Recombinant DNA"). It is possible to recombine DNA in this way because it is ubiquitous among all living beings; genes from an organism in one taxonomic class can be spliced with the genes of an organism in another. That we can engineer sequences that produce novel phenotypes—that is, an organism's observable characteristics—means that the set of all possible genetic sequences can be parsed into categories finer than (1) sequences that result in a phenotype, and (2) those that do not result in a phenotype. Within each of these categories, we also have (a) naturally occurring sequences, and (b) artificially occurring sequences—those created with rDNA technology.

By altering the sequence of nucleobase pairs, genetic engineers can convey different information. The function of rDNA technology may be summed up in the following S expression:

An [organism's genetic material] is recombined in order to generate variations in this {organism}.

I have placed the science terms in brackets because these are the words we will attempt to substitute with their analogs in another domain. I use different kinds of brackets to distinguish among sets of analogs. Removing the particularities of science and retaining just the shape of the phenomenon, we render the following T expression:

An [entity's constituent parts] are recombined in order to generate variations in this {entity}.

The distilled expression conjures additional pairs that respectively fit in these two brackets, such as [bits] and {digital information}, [letters] and {words}, [music notes] and {melody}. We see the permutation pattern in each of these examples: different sequences of [*x*] generate variations in {*y*}. There are two aspects at play here—(a) the focus on constituent parts; (b) the results of their resequencing—and both arise across domains. Mukherjee discusses instances of the former:

> Three profoundly destabilizing scientific ideas ricochet through the twentieth century . . . : the atom, the byte, the gene. . . . The most crucial parallel between [them] is conceptual: each represents the irreducible unit—the building block, the basic organizational unit—of a larger whole: the atom, of matter, the byte (or "bit"), of digitized information; the gene, of heredity and biological information. (9–10)

He asks, "Why does this property—being the least divisible unit of a larger form—imbue these particular ideas with such potency and force?" The answer: "Matter, information, and biology are inherently hierarchically organized: understanding that smallest part [of each] is crucial to understanding the whole" (10). As is understanding the implications of how these smallest parts are ordered. It follows that the variation in entities that results from the resequencing of basic organizational units—and the possibilities of/for entities that may result from novel resequencing—would also be of interest and importance.

Mukherjee quotes biologist Hugo de Vries as saying "the whole organic world is the result of innumerable different combinations and permutations of relatively few factors" (10). The precise number of these "relatively few factors" that de Vries references depends on the entity in question. DNA results from recombinations of four factors—the possible arrangements of nucleobase pairs. All nonradioactive matter on Earth is made up of one or more of eighty naturally occurring chemical elements. The written Torah is composed of permutations of twenty-two Hebrew letters. Each of these examples features a different number of recombined factors, but actually all of these entities can be described using *just two* factors: on and off; yes and no; one and zero. In short, pun intended, they can be expressed using the binary code. Mukherjee explains why: "*All* complex information in the natural world can be described or encoded as a summation of discrete parts, containing no more than an 'on' and 'off' state" (10n). According to him, "This theory was most evocatively proposed by the physicist John Wheeler," who said, "Every particle, every field of force, even the space-time continuum itself—derives its function, its meaning, its very existence entirely . . . from answers to yes-or-no questions, binary choices, bits" (qtd. in Mukherjee 10n; ellipsis in original). Gleick quotes George Gamow as stating that "any living organism is determined by 'a long number written in a four-digital system'" (295). The binary code represents the same idea, just in a two-digital system. Describing a sequence of nucleobase pairs using two factors instead of four would be incredibly long-winded—but possible. In fact, we already have a version of this in the Morse code, whereby a grapheme can be represented as a sequence of two factors. The takeaway here is that though diffuse to the point of being unwieldy, a two-digital system is the most irreducible system there is. The form with the potential to most efficiently depict this yes-or-no pattern is binary code.

Avatars of Genetic Engineering in Fiction

In which instances of fictive discourse can we find a process that reflects how sequence matters? Where in fiction is there a figurative manifestation of the purposeful rearrangement of constituent parts? Rearrangement suggests interaction; indeed, we can locate this pattern in interactive fiction—such as choose-your-own-adventure stories—the salient feature of which is the reader's permutative agency. The architecture of such narratives can be conveyed in the F expression,

> A [story's segments] are recombined in order to generate variations in this {story}.

Let us cast about for narratives that encourage the rearrangement of their constituent parts.

Avatar: *Rayuela*

A story known for hinging on the literary device described by this F expression is Cortázar's 1963 novel *Rayuela* [*Hopscotch*]—a quintessential directedly interactive narrative. The novel "specifically invites reader participation in its production" (Simpkins 61). The author encourages the reader to read the novel's chapters—the building blocks, or constituent parts of the narrative—in different configurations.

In the Tablero de Dirección [Table of Instructions], Cortázar explains to the reader: "A su manera este libro es muchos libros, pero sobre todo es dos libros. El lector queda invitado *a elegir* una de las dos posibilidades siguientes" ["In its own way, this book is many books, but above all it is two books. The reader is invited *to choose* one of the two following possibilities"] (3 [my translation]). The first option is to read chapters 1–56 in order, and the second option is to begin with chapter 73 and then follow a different order laid out by the author. A section of the book called "Del Lado de Allá" ["From That Side"] (9) contains chapters 1–56, and a section titled "De Otros Lados" ["From Other Sides"] (443) contains chapters 57–155. The latter set are referred to as "capítulos prescindibles" ["expendable chapters"] (443); "el lector prescindirá sin remordimientos de lo que sigue" ["the reader can skip what follows without remorse"] (3). This novel's setup foregrounds the notion that order matters. As with nucleobase pairs, bits, and graphemes, there exists a limited number of building blocks (in *Rayuela,* the blocks are the 155 chapters) that can be recombined, and each combination (Cortázar suggests two) yields a different building. The rearrangement of chapters shares the conceptual contours of gene permutation, correlating the literary genre of interactive fiction with the science process of genetic engineering. In both cases, changing the sequence of an entity's elemental units results in different "versions" of this entity; variety results from rearrangement, and the particular sequence matters (Trauvitch, "From Ancient Jewish Texts").

The Pattern Recurs!

Rearrangement is not just evoked by the novel's title. It is also embedded *as a science avatar* in a pivotal juncture of the novel. In chapter 62 we learn that the author-character Morelli had been writing a book, a summary of which mentions "Un sueco trabaja en una teoría química del pensamiento" ["A Swede is working on a chemical theory of thought"] (Cortázar 454 [Rabassa 361)]. A superscript directs us to a footnote that is as long as the chapter proper, presented as an excerpt from the Paris *L'Express* about said Swede, Holger Hyden. Though, as Adriana Verdié notes, *Rayuela* abounds with "pseudo-quotations"—those that "originate from non-existent people, or at the very least, lack the virtue of being references to factually recorded utterances" (140)—Holger Hyden in fact existed, and this particular reference to him is not fictitious. In 1961 *Time* published an article titled "Medicine: The Chemistry of Thought" that Cortázar's footnote reproduces in some parts nearly verbatim. The excerpt and article cover Hyden's talk about his work, which he delivered at a San Francisco conference. Below are the parts I would like to highlight, first in the original Spanish of *Rayuela,* then in Rabassa's English translation, and finally in the *Time* article (my italics throughout):

Rayuela:

> Las funciones superiores del cerebro—la memoria y la facultad de razonar—se explican, para Hyden, por *la forma particular* de las moléculas de proteína que corresponde a cada clase de excitación. Cada neurona del cerebro contiene millones de moléculas de ácidos ribonucleicos diferentes, que se distinguen por la *disposición de sus elementos constituyentes simples,* cada molécula particular de ácido ribonucleico (RNA) corresponde a una proteína bien definida, a la manera como una llave se adapta exactamente a una cerradura. Los ácidos nucleicos dictan a la neurona la forma de la molécula de proteína que va a formar. Esas moléculas son, según los investigadores suecos, *la traducción química de los pensamientos.*
>
> La memoria correspondería, pues, a la *ordenación* de las moléculas de ácidos nucleicos en el Cerebro [. . .]
>
> La riqueza, la variedad del pensamiento se explican por el hecho de que un cerebro medio contiene unos diez mil millones de neuronas, cada una de las cuales encierra varios millones de moléculas de distintos ácidos nucleicos; el número de *combinaciones* posibles es astronómico. (455)

Hopscotch:

> The higher functions of the brain—memory and the reasoning faculties—are explained, according to Hyden, by *the particular form* of the protein molecules which correspond to each type of stimulus. Each neuron in the brain contains millions of different molecules of ribonucleic acid, which are distinguished by the *disposition* of their *basic constituent elements.* Each molecule of ribonucleic acid (RNA) corresponds to a well-defined protein, the way a key is perfectly adapted to a lock. The nucleic acids tell the neuron the makeup of the protein molecule it is to form. According to the Swedish researchers, these molecules are *the chemical translation of thoughts.*
>
> Memory would correspond, therefore, to the *ordering* in the brain of the nucleic acid molecules [. . .]

> The richness and variety of thought is explained by the fact that an average brain contains some ten thousand million neurons, each of which contains in turn several million molecules of various nucleic acids; the number of possible *combinations* is astronomical. (361–62)

Time:

> The higher brain functions of memory and reasoning, Hyden hypothesizes, are achieved by the *way* the neuron alters the protein it forms. Each neuron contains millions of molecules of ribonucleic acid (RNA). Each of these molecules is chemically keyed by the *arrangement* of its internal *building blocks*. These molecules dictate, in accordance with their keys, the nature of the proteins that the neuron forms, in cooperation with the glial cells. The modified proteins are *the chemical representations of thoughts*.
>
> Memory is thus the imprinting of a code on RNA molecules in millions of cells [. . .] The average human brain has ten billion neurons, so the number of possible *permutations* is astronomical.

Proposing that *Rayuela* enshrines Hyden's legacy, scientists Guillermo Delgado and Bruno Estañol summarize this excerpt as describing Hyden's "theories on the chemical nature of mental processes" (409). The core of Hyden's hypothesis is this: Memory corresponds to "the *ordering* in the brain of the nucleic acid molecules"; "these molecules are the chemical translation of thoughts" (Rabassa 362; emphasis added). Otherwise stated, it is the specific *arrangement* of RNA molecules that represents a particular thought. Nested in chapter 62 of *Rayuela*—a novel famous for foregrounding the rearrangement of its elemental units, and which therefore serves as a literary avatar of the permutation pattern—is *another* avatar of the permutation pattern, one that this time manifests in the domain of biochemistry! In both cases, the permutation of constituent elements—whether chapters or nucleic acid molecules—dictates the nature of the outcome.

The reason I view this footnote as a material recursion rather than as a coincidence is that 62 is not your garden-variety chapter among *Rayuela*'s 155. It is the chapter that definitively sets up Morelli as Cortázar's alter ego: an author attempting to write an experimental, subversive novel. There's more: After *Rayuela*, Cortázar wrote a novel titled *62 / Modelo para armar* [*62: A Model Kit*] (1968). The novel *62* literally "derives from chapter 62 of *Hopscotch* and represents his novelistic answer to Oliveira's search for alternatives to that fabricated reality criticized in the earlier novel. In chapter 62

Morelli sketches the outline of 'a book he had been planning but that never got beyond a few scattered notes.' . . . *62: A Model Kit* is the implementation of this attempt" (Alazraki, "Toward the Last Square" 19).[1] Therefore, not only does the novel *62* stem from *Hopscotch*'s chapter 62; the later novel's introductory paragraphs suggest that the reader turn to 62 (the chapter) to render *62* (the novel) more understandable. In these introductory paragraphs Cortázar refers to "las intenciones esbozadas un día en los párrafos finales del capítulo 62 de *Rayuela,* que explican el título de este libro y quizá se realizan en su curso" ["the intentions sketched out one day past in the final paragraphs of Chapter 62 of *Hopscotch,* which explains the title of this book, and perhaps those intentions will be fulfilled in the course of it"] (6 [Rabassa 3]). Chapter 62 stands out among its fellow 154 chapters in a major way: It is the only one that spawns a whole other novel.

Lucille Kerr brings all this full circle in connection with the footnote about Hyden:

> We recall that Morelli, himself an elusive character in *Hopscotch,* posits the foundation for a future book in which psychological causality would cease to govern characters' actions. . . . What may seem most revolutionary is that Morelli's literary theory is supposedly based on the research of a Swiss [*sic*] neurobiologist who proposes that human behavior is caused by chemical changes in the brain rather than by psychological motivation. (230)

Thus, Hyden's research, footnoted in chapter 62, forms the basis of Morelli's literary theory, which in turn forms the basis of the novel *62*. For all these reasons, chapter 62 is unique among the 155 chapters of *Rayuela,* which supports my reading that the permutation pattern's recursion—a foregrounding of elemental units' rearrangement within a foregrounding of elemental units' rearrangement—is meaningful rather than coincidental.

Consider that if the reader follows Cortázar's instructions for reading *Rayuela*'s "second book," they will upon hopscotching to chapter 62 learn for the first time plainly what Morelli is up to: "En un tiempo Morelli había pensado un libro que se quedó en notas sueltas" ["At one time Morelli had been planning a book that never got beyond a few scattered notes"] (454 [Rabassa

1. Elsewhere, Jaime Alazraki observes the following about *62*: "Las primeras treinta páginas definen un ideograma que el resto del texto descifra o intenta descifrar. Aquí están dados todos los ingredientes que, como una receta de cocina, el texto luego combina y mezcla" ("Novela calidoscopio" 157) ["The first thirty pages define an ideogram (symbol) that the rest of the text deciphers or attempts to decipher. Here are given all the ingredients that, like a cooking recipe, the text then combines and mixes" (Google Translate)]. *62* thus serves as another, though structurally different, avatar of the permutation novum.

361]). Many of these notes are among the "expendable chapters" of *Rayuela.* Chapter 62, then, is the very note where Cortázar unveils the embedded fabula that resembles the frame fabula—the novel made up of rearranged notes.[2] And *further* embedded in the embedded fabula is Hyden's hypothesis on the chemical nature of thought, an avatar of the pattern of both the frame and the nested fabulae. Buried in a footnote, as though mentioned offhand, is a description of a structure that parallels both that of the novel within the novel and that of the novel itself; the permutation pattern reappears at three narrative levels.

And so, the decisive trope (in the sense of the Greek τρόπος [tropos], meaning "turn") in the maze of *Rayuela* contains the map to the maze, which also serves as its cipher. The connection is submerged in a footnote, but its significance is telegraphed by its placement in the kernel chapter. This placement suggests that rather than a massive fluke, this third, recursive avatar of the permutation pattern is a deliberate further step into the abyss.

Hyden's hypothesis about the chemical nature of thought is mentioned only one other time in *Rayuela,* in another chapter scholars view as key; consider the following three observations: "Chapter 99 is a crucial chapter in which the characters discuss the role of the writer, literary language and language as such with Morelli's notes at the center of the conversation" (Rodriguez 10); "in Chapter 99 the characters discuss themselves as characters and directly address the author. In fact, they debate his intentions and views about the novel, specifically Cortázar's use of language" (Bartlett 31); "A dismantling of language and genre culminates . . . in a group of chapters (particularly chapter 99) which is a kind of Morelli/Cortázar 'manifesto'" (Jordan 228).

In chapter 99, Oliveira tells the Serpent Club—a group of friends with whom he'd have intellectual discussions, listen to music—that "un sueco acaba de lanzar una teoría muy vistosa sobre la química cerebral. Pensar es el resultado de la interacción de unos ácidos. . . . *Acido, ergo sum.* . . . Ya ves como el *cogito,* la Operación Humana por excelencia, se sitúa hoy en una región bastante vaga, entre electromagnética y química" ["a Swede has just put forth a very impressive theory on cerebral chemistry. Thinking is the result of the interaction of certain acids. . . . *Acido, ergo sum.* . . . You can see then, how the *cogito,* that Human Operation *par excellence,* is nowadays located in a rather vague region, somewhere between electromagnetism and chemistry"] (562 [Rabassa 448]). Here, as in chapter 62, "la causalidad psicológica"

2. As we have seen, while the embedded (that is, Morelli's) novel is a reflection of *Rayuela,* it is fully realized in 62.

["psychological causality"] (456 [363]) is distinguished from "flujos secretos de la materia viva. . . . ese flujo de la materia animada" ["the secret flow of living matter. . . . that flow of animated matter"] (455, 456 [362]); both Morelli and Cortázar want to write characters in their respective books that align with the latter.

Oliveira continues, "Ahí va tu *cogito,* eslabón del vertiginoso flujo de fuerzas cuyos peldaños en 1950 se llaman *inter alia* impulsos eléctricos, moléculas, átomos, neutrones, protones, potirones [*sic*], microbotones, isótopos radiactivos, pizcas de cinabrio, rayos cósmicos: *Words, words, words, Hamlet,* acto segundo, creo" ["There goes your *cogito,* a link in the dizzy flow of forces whose steps are called in 1950, *inter alia,* electrical impulses, molecules, atoms, neutrons, protons, positrons, microbitons, radioactive isotopes, grains of cinnabar, cosmic rays: Words, words, words, *Hamlet,* Act Two, I think"] (562–63 [448]). This quotation features two fascinating components. The first is a list of scientific entities, most of which, it turns out, are elemental units of their respective milieus (and one of which—"microbitons"—seems to be made up). It is these flows of elemental units that are the locations and vehicles of thought, Oliveira suggests, supporting Morelli's note in chapter 62. Here once again Cortázar plays up all manner of building block—which, as we know, he is very interested in having the reader rearrange.

The second fascinating component is that the list of elemental units culminates in Hamlet's famous "Words, words, words" (Shakespeare 694). Shakespeare scholar James Sutton interprets this phrase as Hamlet's metatheatricality and his complication of language[3]—which parallel Cortázar's own use of metafictionality and concern with the limitations of language. We can read this quotation as one more way Cortázar brings our attention to recursive layers in his text. Another, or additional, way to read Oliveira's use of "Words, words, words" is as placeholders in a continuation of the elemental-units list. Indeed, words are themselves elemental units of sentences that are rearranged to relay different meanings. Such a reading of this phrase's inclusion suggests yet a further embedding—and foregrounding—of the permutation pattern in *Rayuela.*

Avatar: *Malas hierbas*

A more recent example of an interactive narrative is Cabiya's 2010 novel *Malas hierbas—Wicked Weeds: A Zombie Novel* in Jessica Powell's 2016 translation.

3. James Sutton, personal communication with the author, May 2026.

Dafne Duchesne-Sotomayor describes the book as "estructurada con un artefacto hibrido, producto del ensamblaje de diferentes tipos de textos y del entrecruce de varios narradores" ["structured as a hybrid artifact, the product of the assembly of different types of texts and the interweaving of several narrators"] (14 [my translation]). The volume is presented as the scrapbook of character and sometime-narrator Dr. Isadore Bellamy, which Emily Maguire refers to as her "meta-narratival exploration" (11). It "includes not only the narratives that Isadore gathers on her own but also the many literary and cinematic references that she collects and analyzes" (10). A formal affordance of the scrapbook that is relevant to our purposes is that a multiplicity of sources could suggest a multiplicity in the shapes information takes and in the perspectives that impart it. For example, information in the form of a police report is quite different from information in the form of a glossary, and both differ greatly from information in the form of ruminations. And the author of each narrative will take on a different voice, as befitting the form. Of course, in truth, the different types of texts that compose *Malas hierbas* are all penned by Cabiya. However, the presentation of the texts' provenance as various is meant to lend to the reader a sense that the volume in their hands is multiformal and multivocal.

Malas hierbas's constituent parts are like *Rayuela*'s constituent parts in that the components of both sets are presented as rearrangeable. And similarly to *Rayuela, Malas hierbas* is prefaced with instructions for the reader to pursue these rearrangements, implying that the actual (published) arrangement depicts just one option. Right after *Malas hierbas*'s page of epigraphs but before the table of contents, we find a section titled "Advertencia" ["Warning"], with the author's name in the byline. I reproduce a large chunk of it here so as to convey the warning's distinctive tone and the foreboding mood it establishes, a mood that is significant in terms of the reader's subsequent approach to the novel. Cabiya explains that the volume you hold in your hands contains "cuatro categorías textuales" ["four textual categories"], and notes,

> El manuscrito ha sido clasificado y segregado a la luz de esos lineamientos, *pero solo en el índice.* El cuerpo del libro repite y respeta el orden en la que aparecen en el original.
>
> . . . Por eso advertimos al lector diligente, ése que leerá en el orden riguroso de la paginación: de nada servirá su meticulosidad y celo, pues siguiendo el orden ingresa al caos de todas formas.
>
> . . . Concluirán otros que, entonces, lo mejor es usar la brújula que provee el índice y leer el libro en el orden establecido por las categorías. Y aunque . . . los capítulos están desperdigados y entremezclados . . . no han de

> temer, puesto que el índice siempre los llevará a buen puerto didiéndoles qué va con qué. Leerán todas las cosas que son de un color y una forma primero, y las que son de otro color y otra forma después, y así sucesivamente, obedeciendo la regla de que cada oveja acompaña a su pareja. Esta conveniencia, no obstante, podría ser fatal.
>
> [The manuscript has been classified and divided in accordance with these criteria of demarcation, *but only in the table of contents.* The main body of the volume respects and maintains the order in which they appear in the original.
>
> . . . And so we issue a warning to the diligent reader, the one who will read in rigorous compliance with the page numbers: Your meticulousness and zeal will get you nowhere, for even following that path you will wind up in chaos.
>
> . . . Others will conclude then, that the best approach is to use the compass provided by the table of contents and to read the book in the order established by the categories. And while . . . the chapters are scattered and intermingled . . . these readers should fear not, for the table of contents will always deliver them to a safe harbor, telling them what goes with what. They will read all the things of one color and shape first, and the things of a different color and shape next, and so on, obeying the rule of like for like. This convenience, however, could be lethal. (vii–viii)]

Cabiya explains that readers who take the second option will not read the volume in the manner intended by its compiler, Dr. Bellamy, "Con funestas consecuencias" ["With tragic consequences"] ([viii]). The warning therefore presents two possible pathways: Read the book in an orderly fashion, according to kind, or take the chaotic route consigned by Dr. Bellamy. This seemingly extradiegetic notice is issued because the reader's experience of the text will in large part depend on which of the two routes the reader takes: Though at the end you will be befuddled regardless of the order you choose, the path you choose makes a difference.

Here's why: Taking the second path—using "la brújula que provee el índice" ["the compass provided by the table of contents"] and reading according to kind—forces the reader to experience the novel as realist rather than speculative fiction. If one reads the five chapters that are categorized as records first, one learns from the outset a scientific explanation for the unnamed protagonist's feelings of zombieness: He suffers from Cotard's syndrome. According to the transcript of Dr. Bellamy's interrogation, Detective Reynolds Rivera Sagardí explains, "La desconexión emocional afecta el reconocimiento de absolutamente todo . . . *incluyéndose a sí mismo.* ¿La historia que inventa el cerebro para explicar la situación? "Debo estar muerto y putrefacto". . . . Todo lo veía según la manera de su fantasía" ["The emotional disconnect affects the recognition of absolutely everything, *including one's own self.* The story that the brain invents to explain the situation? "I must be dead and rotting". . . . He saw everything through the lens of his fantasy"] (237 [153, 154]). Encountering the revelation so early in the narrative obviates an impetus on the part of the reader to suspend disbelief and cultivate a speculative lens. This reframes the presentation of the story and the reading experience.

On the other hand, one who reads according to the order of page numbers discovers the scientific explanation only at the story's conclusion. Let us rewind to the beginning of this reader's process: Alerted to the genre by the book's subtitle, "A Zombie Novel," the reader already enters the storyworld ready to suspend disbelief.[4] Such a reading lens provides magical, fantastical, gothic hues as needed in order for the story to make generic sense. Indeed, in her review of *Malas hierbas,* Lizabeth Paravisini-Gebert recognizes the subtitle as the key that unlocks a range of generic possibilities for the book:

> Cabiya, in subtitling *Wicked Weeds* a "zombie novel," opts for a genre definition that allows him the broadest possible canvas from which to explore . . . the meta-narrative aspects of the zombie tale. . . . *Wicked Weeds* . . . explores the full range of narrative possibilities open to the various subgenres we associate with spec-fic—science fiction, *noir,* dystopia, fantasy, horror, and the supernatural, among them—all of which Cabiya deploys in the telling of his story. (305)

Since the reader reading according to the order of the page numbers comes to know the pharmaceutical executive by way of the exec's own perspective, they must at the conclusion at least question, if not revise, their characterization and framing of the story. The sequence of the story's constituent parts determines the nature of the story and one's understanding thereof.

The Pattern Recurs Here, Too!

The unnamed pharmaceutical executive's mission is to find an arrangement of herbs and weeds that will cure him of his zombieness, as he writes, to "detectar mejor la combinación de factores que operó el fugaz milagro [de sentirse

4. The original Spanish edition does not have a subtitle; there is no paratextual alerting.

vivo] y descubrir la existencia de una formula que me permita completar el proceso de resucitación" ["detect the combination of factors that produced the fleeting miracle [of feeling alive] and discover the existence of a formula that would allow me to complete the process of resuscitation"] (20 [3]). The exec divulges that his decision to study pharmacology and chemistry was in pursuit of this mission: "¿Existe para un zombi una mejor manera de aprovechar su perpetuidad de la materia, los poderes y limites combinatorios de los diferentes elementos, y las dotes curativas de un sinnúmero de plantas y animales, hasta dar con la cura para su dolencia y la de otros muchos?" ["Is there a better way for a zombie to take advantage of his life in perpetuity than by dedicating it to the study of matter, the combined powers and limitations of various elements, and the curative endowments of countless plants and animals in the hope of at least hitting upon the cure for his and so many others' affliction?"] (22 [5]).

Dr. Bellamy as well, albeit for different reasons, is on a quest to understand the attributes of herbs and weeds; the various types of texts in her scrapbook reveal that she "is also interested in how zombies are made. . . . Her search . . . runs parallel to the zombie scientist narrator's search for his own humanity" (Maguire 8). Also like the exec, Isadore describes her goals in terms of her choice to pursue her particular field of study, ethnobotany:

> Los alquimistas registraron los movimientos del espíritu humano y su eventual transformación (la piedra filosofal) observando las propiedades de *los elementos y sus diferentes combinaciones,* revelando, al mismo tiempo, la secreta composición del mundo. Yo haría lo mismo con las hierbas . . .
>
> [Alchemists recorded the movements of the human spirit and its eventual transformation (the philosopher's stone) by observing the properties of *the elements and their different combinations*—revealing, at the same time, the secret composition of the world. I would do the same with the plants . . .] (72, emphasis added [40, emphasis added])

Seeking combinations and recombinations is thus the objective of both protagonists. Their pursuit is mirrored in the novel's structural conceit: Both pursuit and conceit turn on rearrangement.

This sounds familiar because we just mapped the very same pattern's recursion in *Rayuela,* tracing the structural conceit of rearrangement across three narrative levels. In both *Rayuela* and *Malas hierbas,* embedded texts (Hyden's hypothesis; the exec's and Isadore's research) are *science* avatars of the permutation pattern, which parallel the novels' forms as *literary* avatars of the permutation pattern.

Mapping Entailments of Permutation

In the introduction, I argue that metaphorical entailments help further enhance comprehension of a given concept in the target domain. Our permutation fi-sci analogy avails us of at least two such entailments.

Sequence as Signpost

The observation about *Malas hierbas*'s genre alert puts us on a local train that chugs along a little detour. On this route I will adapt Dorrit Cohn's discussion of fictionality signposts and Nielsen et al.'s thesis about the assumption of fictionality to introduce an analysis of *genre* signposts and the assumptions *they* entail.

Cohn argues against John Searle's view that "there is no textual property, syntactic or semantic, that will identify a text as a work of fiction" (qtd. in Cohn, "Signposts of Fictionality" 784). She identifies three signposts of fictionality—textual properties—that "point to the differential nature of fiction" (800) vis-à-vis historical narratives: "the synchronic bi-level (story/discourse) model . . . ; the dependence of certain prominent narratives modes . . . on the constitutional freedom of fiction from referential constraints; and the doubling of the narrative instance into author and narrator" (800).[5]

The "Novel" in *Wicked Weeds*' subtitle leaves little doubt as to the fictionality of the text. However, the "Zombie" modifier signals something additional: the *type* of fiction the reader is about to encounter. The subtitle "A Zombie Novel" announces from the outset that the fiction likely belongs to the fantastical genre. I use the qualification "likely" because the subtitle does not set things in stone. The "warning" you read within a couple turns of the page asks you to choose one of two courses, and it is the course you choose that fixes your fate. If you select the first path and read according to the order of page numbers, you will not reach the scientific explanation until the end and can use the "Zombie" clue in the subtitle to approach the novel as one in the fantasy or speculative genre; if you choose the second path and read according to kind, you will very early on discover the scientific explanation for the

5. Nielsen et al., on the other hand, contend that "it is wiser to talk about degrees of fictionality rather than the distinction of fiction," noting that "a skillful author of nonfictive discourse may take any technique that is currently assumed to be exclusive to fiction and make it function to nonfictive ends" (67). Still, they agree that "some techniques can contextually signal fictive intent" (66), and it is this signaling that interests me here.

pharmaceutical executive's perspective, and approach the novel as one in the realism genre.[6]

As Nielsen et al. note, "The assumption that a text or passage is intended fictively radically alters our reception of it. . . . The assumption of fictionality . . . changes our interpretive activity and its outcomes" (67, 68). If we substitute "fictionality" with "genre" we find that signposts of genre similarly radically alter our reception of a text and change our interpretive activity and its outcomes. Since the only difference between the two paths laid out in the warning is the order according to which the reader consumes the text, we can conclude that sequence—permutation—can function as a generic signpost. The sequence signals to the reader what kind of text they are reading, which decides the kind of interpretive activity they adopt. Applying this back to the permutation pattern in science, we can see that it is indeed analogous: The sequence of building blocks—nucleobase pairs; ones and zeros; chemical elements; graphemes—determines what "genre" (sort) of organism; information; matter; text is built.

Directed Rearrangement

Despite the similarity in terms, *recombinant DNA* and *genetic recombination* are different processes. Table 6 places the differences in context. With which column does the fi-sci analogy between resequencing story segments and genetic engineering align? Beginning with the table's top row, we find that the fi of the fi-sci analogy—interactive fiction—contains the answer in its name. The "interactive" in "interactive fiction" indicates that to be realized, narratives in this genre require outside (reader) participation; the "fiction"—and the very existence of instructions for the reader—conveys artifice. Genetic mutation and recombination occur naturally, without lab involvement. Therefore, the kind of genetic variation we're looking at is artificial genetic variation, also called genetic engineering, or recombinant DNA. Moving on to the table's second row, we ask whether heterologous or homologous rDNA better aligns with the pattern under discussion. When setting up the flowchart expressions, we determined that the things being rearranged are the building blocks—in science that's the DNA; in the two novels we've analyzed that's the narrative's individual units, or chapters. When interacting with interactive fiction, the

6. At least globally, that is. There are sections in the novel that certainly depict local fantasy, but since these sections form an inner fabula, their fantastical nature does not necessarily seep out into the outer fabula, which can therefore maintain global realism. For a discussion of global vs. local fictionality (rather than genre), see Nielsen et al. 67.

TABLE 6. Genetic variation

Natural Genetic Variation		Artificial Genetic Variation (recombinant DNA)	
a. Mutation	b. Recombination	c. Heterologous	d. Homologous
"All types of changes in DNA nucleotide sequence, except . . .	. . . those due to the recombination of fragments of genetic material exchanged between paired DNA strands" (Merlin).	The combination of DNA sequences from two or more different organisms or species	The rearrangement of DNA sequences from a single organism in new configurations

reader is not asked to combine the book's chapters with text that is external to the book (another organism); rather, they are asked to rearrange chapters that all come from the one book (a single organism). Thus, our fi-sci analogy aligns with column (d), homologous rDNA.

Another difference between natural and artificial genetic variation lies in the role that chance plays (or does not play) in each process. Natural genetic variation comes from two sources: mutation and recombination. "Both sources of genetic variation are considered unpredictable phenomena as to the time they occur, the genes they affect, and the individual organisms concerned" (Merlin). Though the rate of random mutations may be affected by environmental factors, the basic level of natural genetic variation—whether as mutation or as recombination—is random.[7] In artificial genetic variation, however, chance does not come into play; the very adjective "artificial" implies the purposeful interaction of an outside actor (e.g., a lab technician), and therefore obviates a role for chance in the recombination.

It is the role of chance—in genetics and in fiction—that affords this fi-sci analogy another metaphorical entailment. *Rayuela* is sometimes characterized as aleatory; aleatory writing involves "an element of randomness either in composition, as in automatic writing and the cut-up, or in the reader's selection and ordering of written fragment" ("Aleatory"). In his analysis of play in fiction, Allen Thiher spotlights *Rayuela* as "a postmodern *locus classicus* of a work that purports to offer an aleatory structure . . . a work whose game title also emphasizes the rule-bound nature of this narration that can be played in more than one way" (168–69); Verdié, in her reading of *Rayuela*'s musical elements, notes that its "structure, similar to the hopscotch game that gives the novel its title, can be likened to an aleatoric music composition, where the musician's re-creation of the modular form within each performance parallels the alternate reading paths chosen by the engaged reader of *Rayuela*" (135).

7. Joan Slonczewski, personal communication with the author, September 2024.

Both Thiher and Verdié suggest that the novel calls on the reader to apply randomness to how they order or consume the text.

But Thiher's "purports" is telling. He explains, "The second order of reading is a prescribed order and is really laid out with all the rigor of the normal 'linear' order offered by the unfolding of pages. The second order appears aleatory only through contrast with the received order of reading given by the book's format. . . . Finally, this novel is not really an actualization of the aleatory" (169). Steven Boldy similarly notes that *Rayuela* "is not a totally open, aleatory novel, nor, as many detractors and enthusers agree, is everything left to the reader" (31)—or to chance, I would add. Though the reader can proceed through *Rayuela*—and any narrative, for that matter—by flipping a coin or rolling dice at the end of each narrative unit (be it a chapter, a page, etc.),[8] this is not exactly what the instructions nudge them to do.

Thiher proposes that "the concrete actualization of the aleatory in fiction" is realized in Raymond Federman's *Take It or Leave It* (1976) and Marc Saporta's *Composition No. 1* (1962) (170). So what is the difference between these two texts on the one hand, and *Rayuela* and *Malas hierbas* on the other, all four of which "invit[e] [their] own decentering" (Thiher 171)? *Take It or Leave It* and *Composition No. 1* ask the reader to shuffle the books' narrative units according to any order, or, one might say, according to chance. But *Rayuela* and *Malas hierbas,* while certainly implying in their instructions that there is no one right order in which one should read their narratives' units, still offer very specific options—particularly, *two* orderings above all. Thus, per their respective paratextual instructions, the former two texts are aleatory in a way the latter two are not.

The metaphorical entailment here is as follows: Just as per the instructions in *Rayuela* and *Malas hierbas* our choice of reading order is not left to chance, so rDNA—as opposed to natural genetic variation—involves directed rearrangement of DNA sequences. This suggests that rather than designating our fiction examples as interactive fiction, to be accurate we should specify that what we're setting up in parallel to genetic engineering is the resequencing of story segments in *directed* interactive fiction.

Any Order?

A corollary of this metaphorical entailment is that in directed interactive fiction such as *Rayuela* and *Malas hierbas* not all possible directions—whether

8. Here we are talking about self-contained narrative units, each of which makes semantic sense on its own. See Thiher 170–71 for a discussion about the much more difficult prospect of shuffling at *other* levels of writing—"at the level of micro-units such as phonemes or syntax."

or not instructed by an author—would make the same level of narrative sense: It is not the case that *any* order goes—in science or in certain fiction. Even back when atomism—"a term used of theories that posit the existence of small indivisible particles as the ultimate components of matter" (Osborne)—was accepted as an explanation of nature, philosophers held that not all combinations of elementary units are possible. In *On the Nature of Things* (50 BCE), Titus Lucretius Carus writes that "there are first-beginnings common to many things, yet they can exist with sums different from one another: so that the human race and corn and glad trees are rightly said to be created of different particles." Lucretius cautions, however,

> And yet we must not think that all particles can be linked together in all ways, for you would see monsters created everywhere, forms coming to being half man, half beast, and sometimes tall branches growing out from a living body, and many limbs of land-beasts linked with beasts of the sea, and nature too throughout the lands, that are the parents of all things, feeding Chimaeras breathing flame from their noisome mouths. (89)

Fantastical imaginings and obsolete theory aside, the core sentiment stands: In biology and in directed interactive fiction like the novels we've explored, not all permutations yield viable results.[9]

Consider that most of the books in Borges's "La biblioteca de Babel" ["The Library of Babel"] amount to nonsense—strings of graphemes that bear no meaning.[10] There is an analog in modern biology: Not all sequences result in a phenotype. Certain combinations of the four nucleobase pairs do not form a functioning organism. Padma Nambisan worries, not unlike Lucretius, that

---

9. "Viable" is to be taken in varying senses depending on the constituent parts in question. "Viable" means a functioning organism in biology, realizable matter in chemistry, a text that makes linguistic sense, and so on.

10. Though it is important to note that Borges's narrator explains why this is not actually true: "En efecto, la Biblioteca incluye todas las estructuras verbales, todas las variaciones que permiten los veinticinco símbolos ortográficos, pero no un solo disparate absolute" ["For while the Library contains all verbal structures, all the variations allowed by the twenty-five orthographic symbols, it includes not a single absolute piece of nonsense"]. Here's why: "proposiciones, a primera vista incoherentes, sin duda son capaces due una justificación criptográfica o alegórica; esa justificación es verbal y, *ex hypothesi*, ya figura en la Biblioteca. No puedo combiner unos caracteres . . . que la divina Biblioteca no haya previsto y que en alguna de sus lenguas secretas no encierren un terrible sentido" ["phrases, at first apparently incoherent, are undoubtedly susceptible to cryptographic or allegorical 'reading'; that reading, that justification of the words' order and existence, is itself verbal, and, *ex hypothesi*, already contained somewhere in the Library. There is no combination of characters one can make . . . that the divine Library has not foreseen and that in one or more of its secret tongues does not hide a terrible significance"] (470 [Hurley 117]). Given the unimaginable size of this Library, however, the possibility that such significance will be discovered is infinitesimal.

certain combinations could "creat[e] monsters defying the definition of 'life'" (86). And there is evidently an analog in some fiction: Not all rearrangements of certain texts' segments yield a story that (a) makes temporal and/or narrative sense, or (b) differs sufficiently from the originally presented order to make any interpretive difference.

Throughout this section I've qualified "fiction" and "texts" with "some" and "certain" because, as we saw earlier, there do exist interactive narratives that call for and encourage any and all segment permutations. In these narratives, such as *Take It or Leave It* and *Composition No. 1*, any order does go. What this says about the metaphorical entailment corollary is that it is partial, rather than absolute. That is, while in biology the statement that not every sequence leads to viable results is always true, in fiction the statement is sometimes true (maybe even many times, but not always.) This caveat reveals something interesting about what maps and what does not. In the introduction I related the notions of metaphorical entailment and the invariance principle: The former pertains to additional properties of the source that can be mapped onto corresponding properties of the target; the latter pertains to those properties that do not map over. What we see now is that even given a metaphorical entailment, there can be aspects of it that map and aspects of it that do not—at least not entirely. An additional property of resequencing story segments (the source) that can be mapped onto corresponding properties of genetic engineering (the target) is that rearrangement is directed, not left to chance. But within this entailment, the corollary of "any order?" reveals the invariance principle: The notion that not every rearrangement is viable pertains to biology but not to *all* fiction. This kind of gatekeeping with entailments and invariance helps us avoid stretching analogies beyond where they are helpful.

In concluding this discussion, let us remember that we are of course free to read any text according to any order; you could begin with your next novel's second-to-last chapter, peruse a few pages in the middle, then read chapter 4. What sets directed interactive narratives apart is that their authors have intentionally invited interplay with the readers, encouraging readers' input—in particular configurations—in the sequence of storytelling. Cortázar and Cabiya are direct in their solicitation, including a dedicated section at the outset of each volume in which they lay out specific sequence options. Directed interactive narratives that operate according to the F expression foreground permutation—the pattern that resequencing story segments has in common with genetic engineering—as an integral component of the meaning-making process.

•

TABLE 7. Permutation flowchart

	Phenomenon	Example	Expression
Science	1 Genetic engineering	2 Editing the sequence of an organism's nucleobase pairs in DNA, thereby encoding different traits.	3 An [organism's genetic material] is recombined in order to generate variations in this {organism}.
Template	—	—	4 An [entity's constituent parts] are recombined in order to generate variations in this {entity}.
Fiction	7 Resequencing story segments	6 *Rayuela*, Julio Cortázar; *Malas hierbas*, Pedro Cabiya	5 A [story's segments] are recombined in order to generate variations in this {story}.

The two metaphorical entailments support the proposition that the permutation pattern applies across the two domains; not only does the pattern hold, but so do some of its offshoots, and this solidifies the pattern's pertinence. This chapter's analysis helps fill out the pattern-mapping flowchart (table 7), which adds to the chorus that this pattern "works," giving us the fi-sci analogy Resequencing Story Segments || Genetic Engineering. Recombinant DNA technology is complex in that we cannot discern it with the naked eye and readily visualize its implications. But having located avatars of its permutation pattern in directed interactive fiction, we stand to improve our conceptualization of this genetic process.

CHAPTER 2

Proliferation—Gametic Reproduction

Chapters 2 and 3 form a duet; they are inextricably but explicably intertwined. Here we'll explore two phenomena in science, each a kind of biological reproduction, that are illuminated, respectively, by two phenomena in fiction, each a kind of literary proliferation.

Reproduction

Having considered the gene as a building block in chapter 1, let's now examine its role as a unit of inheritance and begin by differentiating two kinds—patterns—of reproduction: gametic and clonal. Gametic reproduction, also called sexual reproduction, involves the merging of gametes (haploid germ cells) of different sexes. The offspring of gametic reproduction receives half its chromosomes from each gamete. This offspring's DNA is a convergence of parts from each parent, forming a genetically unique organism. Clonal reproduction, on the other hand, is one form of asexual reproduction. Here each offspring sets out as a genetic replica of the parent and identical to its siblings. Instead of being a product of inheritance convergence, an offspring of clonal reproduction is the product of inheritance iteration.

Genetic variation is "built in to" gametic reproduction, but it happens in clonal reproduction as well. In both kinds of reproduction, variations arise due to random mutations—those that occur in the course of reproduction, and those that occur during the lifecycle of each organism. Epigenetics—changes in gene function without modifications to the DNA sequence and influenced by the organism's environment—also adds to diversity. Because of this even the offspring of clonal organisms can be genetically and/or phenotypically different. Thus, though the extent of variation differs in each kind of reproduction, the offspring of both are subject to it.

There is another characteristic, however, that contrasts gametic and clonal reproduction, and it is the characteristic we will focus on in the fi-sci analogies of this chapter and the next chapter: the location of multiplicity in each process. In gametic reproduction the multiplicity is in *inheritance*: The element distinguished as manifold rather than singular is the provenance of heredity, which is necessarily multiple—generally, there must be *two* parents to produce an offspring.[1] These two parents could have multiple offspring, but multiplicity is not necessary on the offspring side—only on the parent side. In clonal reproduction the provenance of heredity is singular; here multiplicity is located in *iteration*. What's distinguished as manifold is the singular parent's copies. Of course, the offspring of clonal reproduction need not be multiple; the parent might reproduce just once. But the provenance of heredity in clonal reproduction will only ever be singular, so if multiplicity were to appear anywhere, it would be in the iteration of offspring. With these caveats, let's differentiate the characteristic of multiplicity more precisely:

- In gametic reproduction, multiplicity *must* occur on the parent side;
- In clonal reproduction, multiplicity *can only* occur on the offspring side.

There is actually another way multiplicity differs in gametic and clonal reproduction—this time relating to additional senses of the word. In gametic reproduction, "multiplicity" also carries the sense of *difference*: It is not just that the parents must be two—each of the two must be of a different mating type. Here, reproduction is a convergence of differences. In clonal reproduction, on the other hand, "multiplicity" does not carry the sense of difference. It is used in its sense of *quantity*: It pertains to the number of like offspring.[2] Here, reproduction is replication. We can now be even more precise in differentiating multiplicity in each type of proliferation:

- In gametic reproduction, multiplicity *must* occur on the parent side, where the term's sense of *difference* is key;
- In clonal reproduction, multiplicity *can only* occur on the offspring side, where only the term's sense of *quantity* is applicable.

1. With some exceptions, such as self-pollination in the plant kingdom.
2. I credit Katra Byram for pointing out these different senses of "multiplicity."

Replacing the more discipline-neutral terms in the T expression with ones that pertain to transtextuality allows us to render the F expression as,

In [transtextuality] multiplicity lies in {precursor texts}, whose <voices> converge in a new (text).

Avatar: *Adios, Happy Homeland!*

While broadly speaking any text is transtextual, certain narratives are purposely and foundationally so. A narrative paradigmatic of conscious transtextuality—one whose form is wrought from a network of relationships to other texts—is Ana Menéndez's *Adios, Happy Homeland!* (2011). In her review of this book, Amy Letter describes it as a work that "makes many references to the works of others. It is presented in the form of an anthology, with each of the interlinked and intertwined stories being attributed to a different fictional author" (1). Alli Carlisle adds, "Many of the stories are held together by an internal intertextuality, with authors and characters appearing in each other's stories" (166). Some of the fictional authors are fictionalized versions of or allusions to nonfictional authors, such as Alex Carpenter (Alejo Carpentier), Ernesto del Camino (Ernest Hemingway), Joseph Martin (José Martí), and Victoria O'Campo (Victoria Ocampo). Carpenter's "contribution" to the "collection"—"Journey Back to the Seed (*Qué Quieres, Vieja?*)"—is a variation on Carpentier's "Viaje a la semilla" ["Journey Back to the Source"]. The anthologist of this "collection," one Herberto Quain (a version of Herbert Quain, who is a character invented by Jorge Luis Borges), includes in his prologue a footnote citing Charles X. Kinbote; Charles Kinbote is a fictional character, the narrator of Vladimir Nabokov's *Pale Fire* (1962), a novel famous for containing the bulk of its narrative in the endnotes, stylized as "Commentary." The above is just a sample of the intricate, transtextual web Menéndez weaves—the way she funnels a multiplicity of voices that converge in her book.

Among the preceding examples alone we can already identify all five of Genette's types: intertextuality (e.g., Alex Carpenter alludes to Alejo Carpentier); paratext (Herberto Quain's prologue); metatextuality (Charles X. Kinbote evokes *Pale Fire*); hypertextuality ("Journey Back to the Seed [*Qué Quieres, Vieja?*]" as hypertext to the hypotext "Viaje a la semilla"); architextuality (the novel presents as an anthology). What does differentiating these types of relationships between texts do for us? How does it benefit our analysis of *Adios*? I submit that it illustrates just how deep, elaborate, and wide-ranging the novel's drawing on heredity is. It allows us to hear the book say, "my provenance manifests in gales and breezes, from this side and that, in sprinkles and splashes." This sort of broad scope in intensity, direction, and kind foregrounds the extent to which heredity informs and permeates the volume. If we were using "intertextuality" rather than "transtextuality," the subtle and not-so-subtle differences between the types would be lost.

More than being inherently transtextual, *Adios* is a work of fiction that relies on the affordances of fictionality; unlike realist fiction such as *To Kill a Mockingbird*, which is plausible, *Adios* is underpinned by the impossible. Carlisle writes, "The particular fictional universe Menéndez constructs . . . creates its own magical reality with a blend of literary reference, metafiction, and fantasy" (179). In acts of "marvelous metafictional bravado" (Letter 1), Menéndez inflects her brand of transtextuality with self-referentiality galore. The most prominent example is that pertaining to Quain.[5] The scenario it sets up has a fictional character (Quain) created by a different nonfictional author (Borges) take on the role of the actual author (Menéndez)—and who "could be read as her avatar" (Carlisle 166)—in the course of which he (Quain) collects stories by authors that arguably inhabit various ontological strata: Some are straightforwardly made up by Menéndez; others are fictionalized versions of nonfictional ones; one, the "Ana Menéndez" who contributes the ultimate piece to the collection, is a fictionalized version of the actual author, Ana Menéndez. This scenario is fantastical because it is impossible; it could not happen outside of fiction.

The transtextuality coursing through Menéndez's book is not unlike the "transtextuality" suggested by inheritance; the multivocal stories our genomes tell. *Adios, Happy Homeland!* carries a photograph of its literary ancestors in its wallet. Addressing this analogy, [the real] Menéndez remarks, "indeed, in many ways books are living organisms with 'pasts' that bear on their beings."[6]

5. In an excavation of *Adios* we unearth many more allusions to, quotations of, variations on—in short, transtextuality vis-à-vis—the work of Borges. (Ada Ortuzar-Young remarks, "The frequent reliance on apocryphal writers, the motif of the labyrinth and the dizzying effects of illusion and reality show [Menéndez's] debt to the magical realist techniques of Jorge Luis Borges" (166).) The incorporation of Herbert Quain is perhaps the most metafictional at this archeological site.

6. Ana Menéndez, personal communication with the author, June 2020.

Architexture

Among the five types of textual transcendence, I'd like to look further into architextuality—"the relationship between the text and its architext" (Genette, *Architext* 83)—as it pertains to *Adios,* particularly the book's masquerading as an anthology. In what follows I reproduce Genette's detailed characterization of architextuality, then examine how its components apply, or do not apply, to the case of *Adios.*

> The fifth type . . . the most abstract and implicit of all, is *architextuality*. . . . It involves a relationship that is completely silent, articulated at most only by a paratextual mention, which can be titular . . . or most often subtitular . . . but which remains in any case of a purely taxonomic nature. When this relationship is unarticulated, it may be because of a refusal to underscore the obvious or, conversely, an intent to reject or elude any kind of classification. In all cases, however, the text itself is not supposed to know, and consequently not meant to declare, its generic quality. . . . One might even say that determining the generic status of the text is not the business of the text, but that of the reader, or the critic, or the public. Those may well choose to reject the status claimed for the text by the paratext. . . . But the fact that this relationship should be implicit and open to discussion . . . or subject to historical fluctuations . . . in no way diminishes its significance; generic perception is known to guide and determine to a considerable degree the readers' expectations, and thus their reception of the work. (*Palimpsests* 4, 5)

- "a relationship that is completely silent, articulated at most by a paratextual mention"

Adios's classification as an anthology is indeed articulated by a paratext: the prologue attributed to Herberto Quain. There Quain declares, "What follows here is a modest anthology of my life's investigations" (Menéndez 11). But in this book the relationship between the text and the category to which it purportedly belongs is the opposite of silent. For one, the articulation is not a singular occurrence, made once and then never again. In the middle of Menéndez's book are two sections that double (triple?) down on the novel's anthology costume: the message "From: The Poets, To: Herberto Quain, Date: May 23, 1923, Re: Your book," immediately followed by the response, "From: Herberto Quain, To: The Poets, Date: August 9, 1912, Re: Re: Your Book." These intervening sections serve to raise the volume on signaling the book's false architexture.

Moreover, anthological indicators are woven into the very narrative in a web of internal and shared references. In the prologue, Quain mentions Victoria O'Campo's *A Brief History of the Cuban Poets,* which piece appears in the form of a contribution in the book; Laika Almeida and Gertrudis Gómez are two contributing authors who show up as characters in stories contributed by other authors; a hole in the ceiling or firmament features in four of the stories; the words "stars" and "constellations"—particularly Aries and Pisces—"fish," "balloons," "nylon," and "trains," and the themes of flight and escape permeate the volume; the phrase "there are affections of such delicate honesty" closes the contribution by Silas Haslam (another character from Borges) and opens the contribution by Jane Smith.

It is one thing when an author brings up the same ideas several times in a single work, another when a collection of authors do; in a single-author text (SAT) these repetitions are motifs, whereas in an anthology they are internal and shared references that amount to a chorus. When fictional or fictionalized contributors to an ostensible anthology reference the same notions and items—and each other—readers understand them as cohabiting a consistent shared universe. The continuity and community this evokes strengthen the cohesion of a multi-author collection, which imparts to the book that chorus quality. Soon I will return to the difference between the SAT and the anthology and discuss what the signaling of a false architexture might accomplish (pin #1).

- "most often subtitular"

This characteristic does not apply to *Adios,* but as I explore in chapter 1, it does apply to *Malas hierbas,* where the subtitle "A Zombie Novel"—which appears only in the English translation—does interesting work. Speaking of signaling a false architexture, though, we should note that "Novel" in *Malas hierbas'* subtitle—a signal of that book's true category—is at odds with the false generic declaration made in the prefacelike "Warning," according to which *Malas hierbas,* like *Adios,* is a kind of anthology—this time a scrapbook that collects a multitude of voices.

And just like in *Malas hierbas,* incidentally, within the collection that is *Adios* are additional collections. Where in *Malas hierbas* there's an appendix that is a "Glossary of Wicked Weeds," in *Adios* are found the following: "Glossary of Caribbean Winds"; "Zodiac of Loss"; "Adios Happy Homeland: Selected Translations According to Google"; "A Brief History of the Cuban Poets"; "End-less Stories"; "Contributors' Notes." In addition to further burgeoning the book's meta sensibilities, these collections within the collection serve to extend the reach of *Adios*'s transtextual web. Each item in each list is another node in the network, collecting and connecting to ever more precursors.

- "*architextuality*. . . . involves a relationship that remains . . . of a purely taxonomic nature"

Taxonomic, yes, but taxonomic plus. In other words, the taxonomic designation serves a purpose that is integral to *Adios*'s narrative and how the audience perceives it, as discussed above and below.

- "when this relationship is unarticulated, it may be because of a refusal to underscore the obvious or, conversely, an intent to reject or elude any kind of classification"

In the case of *Adios,* the relationship between the text and its architext is not just articulated but foregrounded. Menéndez goes to lengths to establish a classification—a false one.

- "the text itself is not . . . meant to declare, its generic quality. . . . determining the generic status of the text is not the business of the text"

Not so in *Adios,* which declares, or takes on, the specious generic designation of an anthology, one that is moreover crucial to its program. And to ours. To see why, let's consider the etymology of "anthology":

> 1630s, "collection of poetry," from Latin *anthologia,* from Greek *anthologia* "collection of small poems and epigrams by several authors," literally "flower-gathering," from *anthos* "a flower" (see anther) + *logia* "collection, collecting," from *legein* "gather" (see -logy). The modern sense (which emerged in Late Greek) is metaphoric: "flowers" of verse, small poems by various writers gathered together. ("Anthology")

The main markers of an anthology are "collection," "several authors," and "gathered together." The first and the third apply to *Adios*; the second does not (but Menéndez asks the audience to read the book as though it does). Though in different ways from the SAT, the anthology, too, relates to gametic proliferation because it is also the result of drawing on ("gathering together") many voices ("several authors"). If the anthology relates to gametic proliferation, why don't we just look at actual anthologies for the fi component of our present fi-sci analogy? The difference between the SAT and the anthology—to be examined shortly—will serve to answer this question (pin #2).

- "generic perception is known to guide and determine to a considerable degree the readers' expectations, and thus their reception of the work"

In the previous chapter we saw how this is the case in *Malas hierbas.* Readers receive a signal from the subtitle that sets their expectations; then, their generic perception and reception of the work is likely determined by the order in which each reader chooses to read the book's chapters. In *Adios,* it is not a subtitle or permutative agency that determines generic perception. Rather, as we've noted, the author asks readers to take on a false generic perception. To determine how the ask and the perception guide readers' expectations and reception of the work, we would benefit from considering the rhetorical model of audiences. Phelan writes the following about *Adios*:

> The distinctions between narrative, authorial, and actual audiences strike me as a helpful way to get at the complexity of the ontological play Menéndez is engaged in. Authorial audiences pick up on the differences in ontological levels and see Menéndez as the agent behind the play; narrative audiences don't. For example, they view Quain as the author/editor of the volume. Actual audiences (or the subset of them interested in Menéndez's intentions) participate in both audiences.[7]

The differences between the types of audience are introduced and explored in more detail in chapter 4; for our present purposes, it is valuable to realize that for the narrative audience Menéndez's book is an anthology, that is, a collection of texts each written by a different author, while for the authorial audience it is a collection of texts all written by the same author that is masquerading as an anthology. And thanks to this, actual audiences, who participate in both narrative and authorial audiences, are given an additional reason to view the book in terms of biological proliferation. Recall the first reason the actual audiences relate *Adios* to proliferation: The authorial audiences view the book as a being in which the voices of multiple predecessors converge, which is analogous to gametic reproduction, where multiple predecessors converge in an offspring. So that I can describe the additional reason—the one given to the narrative audiences, this time—I will now revisit the two matters I put

7. James Phelan, personal communication with the author, February 2024.

pins in earlier: (1) what the signaling of a false architexture might accomplish, and (2) the differences between the SAT and the anthology.

Matter 2 first. Upon reviewing the etymology of "anthology," I noted that the aspect thereof that does not apply to *Adios*—and all other single-author texts—is the qualification that the book feature several authors. The two other markers fit: *Adios* is indeed a collection, a clutch of texts, some of which were originally published elsewhere and elsewhen (see the book's copyright page); these disparate texts were gathered together in the volume titled *Adios, Happy Homeland!* The quality that precludes *Adios* from being an actual anthology—the fact that it is a single-author text—is the very quality that lets it fit the F expression,

In [transtextuality] multiplicity lies in {precursor texts}, whose <voices> converge in a new (text).

The S expression's analog of the F expression's "text" is "organism." It seems to me that a text that is an organism—a single and singular creature—is a text that has a single author. And in an anthology, which is a text that collects texts by multiple authors (in each of whom precursor voices converge), each of the collected texts is a sibling to the others: As suggested by their collection within the same covers, the texts relate to each other—have something in common. In short, the difference between a SAT and an anthology is that the former is a convergence, and the latter is a group of convergences. The reason anthologies do not fit the parameters of the fi in the present fi-sci analogy is that we seek a convergence of voices in a new, single voice.

With the answer to matter 2 in tow, let us move on to matter 1: what the signaling of a false architexture accomplishes. *Adios* pretends that the convergence in a single voice is in fact an assemblage of multiple voices ("various writers gathered together"). In an anthology, what are otherwise a solitary author's motifs emerge as different authors talking about related things, which amounts to a chorus of validation. But we know that *Adios* is not an anthology, so the signaling of a false architexture accomplishes this: In presenting her work as a collection of multiple voices, Menéndez lays bare her predecessors; she actually highlights even more brightly the truth, which is that all of these precursors' voices converge in hers.

I now return to the reason Menéndez provides the narrative audiences—in which role readers view the book as an anthology—to relate *Adios* to biological proliferation. As an anthology, the book is a group of siblings, each text related to the others by a through line—that thing they have in common that brings them together under one roof / within the same book covers. Now the actual audiences, who participate in both narrative and authorial audiences, have an additional reason to view *Adios* through a biological lens: In the authorial-audience mode the book is an offspring, a being in which the voices of multiple precursors converge (which comes into starker relief by virtue of Menéndez's false generic signaling); in the narrative-audience mode the book is a group of siblings, a cluster of related beings in each of whom voices of multiple precursors converge. Progenitors having multiple offspring is indeed a common aspect of gametic reproduction. Transtextuality in a single-author text is one means of proliferation. An anthology, which groups related single-author texts, thus functioning as extension of gametic reproduction, is another.

•

The matter of architextuality in *Adios* is complicated by the novel's signaling of a false architexture. If so many components of Genette's characterization of architextuality do not neatly apply here, it is not because they are inaccurate. Rather, it is because his characterization does not take into account the special case of a text that seeks to articulate its own generic quality by nondeceptively pretending that it belongs to a category to which it does not in fact belong. Here we are reminded of Gorman's definition of fiction as "intendedly but *non-deceptively* untrue" (163, emphasis added). Menéndez is not trying to trick us into actually believing that *Adios* is an anthology put together by Quain; instead, she is asking us to approach the narrative as though it is such an anthology. In my view, this nondeceptive ask serves to further strengthen the parallel between transtextuality and gametic reproduction.

Mapping Entailments of Proliferation: Convergence

Collections of Convergences

The notion of an anthology as a group of sibling—by which I mean related—texts emerges as a metaphorical entailment. Biological siblings are offspring who have parents in common. An analog in fiction is an anthology of stories whose authors share literary precursors. *Adios* purports to be such an anthology; the aforementioned numerous internal and shared references that bloom throughout the book demonstrate that its authors have literary precursors in common. Interestingly, in their letter to Quain the authors themselves reject their grouping. In "From: The Poets, To: Herberto Quain, Date: May 23, 1923,

Re: Your book," they write, "Because we are such a diverse group of writers, you may imagine that the reasons for asking you to cease and desist are equally numerous. Many among us object, first, to being lumped into that most discredited of categories: *Cuban*" (Menéndez 99). But the reasons they provide suggest their objections pertain to things other than a lack of shared literary precursors:

> Do you intend our work to languish in the special-interest section of the great library? Are you actively working to keep us tied to a single identity?. . . . And now you come, an outsider, to impose on us a doomed structure. . . .
>
> Which brings us to another objection: Who are you? You are neither Cuban nor a poet yourself. . . .
>
> As one of us said, Quain, quién? What gives you the authority? To be done properly, this collection should be put together by a genuine scholar, preferably Cuban himself. . . . we respectfully ask that you leave the Cubans to the Cubans. (99, 100)

In the course of laying out their grounds for objection, the authors actually acknowledge what they have in common: "As you no doubt understand by now, much of our lifework was and continues to be dedicated to the idea of escaping the bonds imposed by others" (99). This would indicate that there is a shared history, which makes it likely there are also shared literary precursors.[8]

Progenitors can have multiple offspring, in each of which the progenitors' DNA converges. Literary precursors can influence multiple different texts, in each of which the precursors' voices converge. Just as offspring can have siblings with whom they share biological predecessors, single-author texts can be grouped with other single-author texts with which they share literary precursors.

Travel

Fiction makes possible intersections of characters and people who could not come across each other outside it. In *Adios,* the old man from Hemingway's *The Old Man and the Sea* (or a version thereof) meets a fictionalized Elián González, a Cuban child at the center of international custody and

8. Those who live under the same roof or within the same book covers can have things in common that are not predecessors: Just as there are anthologies consisting of texts that do not share literary precursors, there are siblings who do not share biological parents.

TABLE 8. Proliferation (convergence) flowchart

	Phenomenon	Example	Expression
Science	1 Gametic reproduction	2 When dolphins reproduce, the offspring receives half of its chromosomes from each progenitor.	3 In [gametic reproduction] multiplicity lies in {progenitors}, whose <DNA> converges in a new (organism).
Template	—	—	4 In [a certain type of proliferation] multiplicity lies in {predecessors}, whose <characteristics> converge in a new (entity).
Fiction	7 Transtextuality	6 *Adios, Happy Homeland!*, Ana Menéndez	5 In [transtextuality] multiplicity lies in {precursor texts}, whose <voices> converge in a new (text).

immigration disputes in 2000. Though the identity of the characters is not plainly stated as such, Menéndez leaves several clues that make it all but certain. The narratives of these two people traveled a certain literary distance (in the sense that each person originally appeared in a different text), at times perhaps through intricate textual pathways (if the way they come together is more convoluted than straightforward)[9] to converge in *Adios*. Said pathways would include other narratives that mention or allude to either; I am not aware of an additional work of fiction that features their meeting.[10] In *Adios*, the old man and "the boy" come together on new ground, so to speak.

In biology, an analog of this journey is made by recessive alleles. These alleles can "lie dormant" for generations, suddenly surfacing in a particular convergence of gametes.[11] A woman and man with brown eyes could have a blue-eyed child as long as both have a blue-eyed ancestor—no matter how long ago (or how long apart from each other) these ancestors lived. In this child the alleles that determine the eye color could be meeting for the first time.

9. Cf. a biology variation: ". . . buscarse por los laberintos más intrincados de la sangre" [". . . seek each other through the most intricate labyrinths of blood"]. (Gabriel García Márquez 470 [Rabassa 422]).

10. But even if such a work exists, the meeting would not be identical to the one described in *Adios*.

11. I use recessive rather than dominant alleles in this analogy because the comparison is between characters "traveling" before they come together and alleles lying dormant for generations (which dominant alleles do not do).

Mapping this additional property of the source (transtextuality) onto a corresponding additional property of the target (gametic reproduction) supports the suggestion that the convergence pattern is indeed applicable across the fiction and the science (table 8 presents the flowchart for the Transtextuality || Gametic Reproduction fi-sci analogy). This metaphorical entailment whereby two entities with different provenances "travel" before meeting in a new creation strengthens the original pattern's claim to legitimacy.

CHAPTER 3

Proliferation—Clonal Reproduction

Duet Fi-Sci Analogy II. Clonal Reproduction: Iterations

Where gametic reproduction takes the pattern of convergence, clonal reproduction takes the pattern of iteration. In this form of asexual proliferation, multiplicity can only be located in copies. Let's convey iteration in biological evolution as the S expression,

> In [clonal reproduction] multiplicity lies in iterations, each of which is a {copy} of a singular <progenitor>.

Replacing science terminology with more "neutral" terminology, we get the following for the T expression:

> In [a certain type of proliferation] multiplicity lies in iterations, each of which is a {version} of a singular <entity>.

To visualize these iterations, we turn to a trembling giant. Pando, Latin for "it spreads," is a quaking aspen that covers over one hundred acres by central Utah's Fish Lake. This tree is a clonal organism: Each of its 47,000 stems, or ramets, is a clone that springs from the same source. "Unlike giant sequoias," explains biologist Michael C. Grant, "each of which is a genetically separate individual, a group of thousands of aspens can actually be a single organism, sharing a root system and a unique set of genes." Thus, what appear as distinct trees are in fact iterations of a single genome. And Pando is *ancient.* The US Department of Agriculture's Forest Service estimates the giant has been around since the end of the last ice age, making it about ten thousand years old. It gets more fabulous: This creature could potentially live forever. Says Grant, "In principle, clones may even be essentially immortal, dying only from disease or the deterioration of the environment rather than from some internal clock."

Friends of Pando is an organization "dedicated and working to educate the public, support research and preservation efforts and inspire stewardship of Pando, the world's largest tree" ("About Us"). Their beautiful website offers views of Pando's [above-ground] extent, and, thanks to the Pando Photographic Survey, an opportunity to "explore Pando in 360°" ("Explore Pando"). But even these visuals do not make the massive clonal organism easy to conceptualize. The entire forest is one tree? With no internal clock? The way this creature behaves in both space and time is so different from what we imagine when we think of an organism, it may as well exist in a parallel universe with alternate physical laws. Friends of Pando observe, "The idea of a tree that spans across 106 acres . . . doesn't fit with our commonsense view of what a tree is, or can be. Pando is a tree that transcends nearly every concept of trees we have today and defies the bounds of every classification system we have for large trees" ("What Is Pando"). We find a similar sentiment in Richard Powers's novel *The Overstory* (2018), when Dr. Patricia Westerford describes her encounter with this very same *Populus tremuloides*: a "great, joined, single clonal creature that looks like a forest. . . . one of the oldest, largest living things on earth. All around her spreads one single male whose genetically identical trunks cover more than a hundred acres. The thing is outlandish, beyond her ability to wrap her head around" (131). Even someone whose expertise is trees finds it difficult to conceptualize an organism at this scale.

One perplexing aspect of such a clonal organism is the implication of this kind of joinedness—interconnectedness—for notions of organism individuality. In his study "Thinking Forests in Ursula K. Le Guin's 'Vaster Than Empires and More Slow' and *The Word for World Is Forest,*" Dion Dobrzynski cites Suzanne Simard's research of the mycorrhizal networks that enable communication among forest organisms and her comparison of this process to that of a neurological network, "suggesting a kind of sentience that emerges from complex, interconnected systems that may even transcend the traditional conception of what an individual organism is" (Dobrzynski 46). Though Dobrzynski and Simard are discussing interconnectedness among a forest's multiple individual trees, the description of transcending the traditional conception of an individual organism is applicable to Pando, a forest made up of just *one* tree.

Peter Godfrey-Smith writes, "The theme of 'individuality' has become a point of contact between biology and philosophy," raising questions such as "What is the basic unit of life or living organization?" (17). And while organisms such as humans and fruit flies "have fairly clear boundaries in space and time," "a move to plants is all it takes for things to get more difficult" (18). Godfrey-Smith explains that one reason for this difficulty is "an uncertain relation between *growth* and *reproduction*. Reproduction is tied to individuality: reproduction is making a *new* individual, while growth is making more of the same" (18). This distinction mirrors the one at the heart of this chapter: Gametic reproduction creates a new individual (convergence), and clonal reproduction is growth that makes "more of the same" (iteration). "Perhaps a biological individual," Godfrey-Smith submits, "is just any object that some part of biology recognizes as worth describing. It might be an organism, a part of an organism, or a larger thing like a colony or ecological system. There are no fundamental or most-real individuals in biology" (19). But while this may explain variety in definitions of an individual organism, the notion that an entire forest is an individual organism is still challenging to fathom.

It isn't that humans cannot conceptualize clones (we can and do); rather, our notion of *tree* or even *organism* does not adhere to a manifestation of such multiplicity, nor to one of size and age on this scale. And so, we find that there are two distinct but related issues at play:

a. That which appears as many discrete entities is actually one organism;
b. That which is massive in space and vast across time is actually one organism.

A trait that applies to both (a) and (b) is continuity: The many that are one are interconnected; the massive and vast that is one persists across space and time. This continuity in both is not readily apparent to the human eye: Interconnection in (a) takes place underground; persistence in (b) is too big and slow. In the present corner of the book we are looking at iteration—issue (a)—which is the focus of this chapter duet's second half. But since the matter of scale is integral to why we may not instinctively conceive the multiplicity in Pando as making up one organism, let us carefully examine (b) as well.

Too Big and Slow

It is no coincidence that (b)'s "too big and slow" sounds a lot like the title of Ursula K. Le Guin's 1971 story: "Vaster Than Empires and More Slow" features continuity via (a) and (b)—both the interconnectedness and the persistence of—a literally outlandish forest. World 4470 is "a dark-green jewel" (189), the vegetation that envelops it "an unmoving giant" (192). Even after flying halfway around the planet, its surveyors realize they are still within the same organism. "It's all one," Osden says. Like Utah's giant, Le Guin's creature is "one big green thought" (212). Harfex's description of the vegetation is spot-on, even as he claims his explanation is impossible:

> The branches, the epiphytic growths, the roots with those nodal junctures between individuals: they must all be capable of transmitting electrochemical impulses. There are not individual plants, then, properly speaking. Even the pollen is part of the linkage, no doubt, a sort of windborne sentience, connecting overseas. But it is not conceivable. That all the biosphere of a planet should be one network of communications, sensitive, irrational, immortal, isolated. (212)

Notwithstanding some hyperbole—quaking aspens do not connect overseas; plant sentience is a subject of research but not currently accepted across the board[1]—Harfex may as well be describing Pando: The 47,000 ramets are all one.

It isn't easy for the characters to describe this planet-organism; "they struggle to come up with words, scientific explanations, and metaphors to deal with 'it'" (Farina 437). Marco Caracciolo notices that Harfex even "resorts to abstract, computational language ('merely a network of processes')" (172). As for the readers, a story like "Vaster" tries to facilitate conceptualization as it "attempts to bridge across scalar levels by blending distinct semantic domains—for instance, by applying notions of agency and subjectivity beyond the human" (Caracciolo 163). Caracciolo points to "the scale variance involved in the upward movement from human-scale organism . . . to planet-scale superorganism" as the culprit of inconceivability (172). But "Vaster" moves beyond conceivability to basic perception; Le Guin explores why organisms

1. See Segundo-Ortin and Calvo; Plebe. See also Dobrzynski: "Controversially, scientists working in a new field they call 'plant neurobiology' (first heralded in a provocative article in the 2006 issue of *Trends in Plant Science*) are calling for plants to be recognised as exhibiting 'intelligent,' and potentially 'sentient' behaviour, despite the absence of a neurological system" (46).

that live at such different spacetime scales may not even be *aware* of each other—much less able to imagine one another. Harfex insists that the vegetation "would not be capable of conceiving of a self-moving, material entity, or responding to one. It could no more become aware of us than we can 'become aware' of Infinity" (Le Guin 208). Mannon adds that to a forest "we might appear as forest fires. Hurricanes. Dangers. What moves quickly is dangerous to a plant. The rootless would be alien, terrible" (208).

The challenge of conceivability also reverberates across *The Overstory*—itself a novel about the interconnectedness and persistence of trees. In its first manifestation, the challenge of conceivability across scales is described in a story about otherworldly creatures:

> Aliens land on Earth. They're little runts, as alien races go. But they metabolize like there's no tomorrow. They zip around like swarms of gnats, too fast to see—so fast that Earth seconds seem to them like years. To them, humans are nothing but sculptures of immobile meat. The foreigners try to communicate, but there's no reply. Finding no signs of intelligent life, they tuck into the frozen statues and start curing them like so much jerky, for the long ride home. (97)

The "frozen statues" are "living things that do not seem alive enough—that live too slowly, we might say, on the 'lower' temporalities" (Goble 226). The challenge of conceivability across scales continues echoing throughout the novel; below is a selection of its pop-ups:

- Prophesying more advanced programming, Neelay describes its digital components as "Living things. . . . So fast, they'll think we're not even here" (107).
- Neelay explains why it took him some time to notice the giant redwood right next to him: "He knows how he missed it: It's too big. Too big to make sense of. Too big to credit as a living thing. . . . And up the trunk runs, straight up, beyond comprehension, an immortal, collective ecosystem—Sempervirens" (197).
- In a therapy session Mimi detects in her patient "thoughts too large and slow to understand" (351).
- Dr. Westerford writes about the wisdom and generosity of trees, adding "Trees have long been trying to reach us. But they speak on frequencies too low for people to hear" (424).
- Asked how he pleads, Adam "blinks, as if he's another form of life altogether and human speech is way too fast to understand" (467).

- Neelay's digital creations—data collectors, surveyors, artificial intelligence, advanced programming referred to as his "learners," his "children," or "bots" (perhaps the "living things" he mentions in the line quoted from p. 107)—"watch and match, encode and see, gather and shape all the world's data so quickly that the knowledge of humans stands still" (487).

The parable itself is repeated again toward the end of the novel: "Aliens land on Earth. They operate on a different scale of time. They zip around so fast that human seconds seem to them as tree years seem to humans" (487).

All of this forms a leitmotif in *The Overstory*, a tale about rescaling, awareness of the Other, and learning to see anew. The "agency of vegetal life . . . usually remains hidden below the relatively coarse levels of human perception due to its diminutive scale and the relative slowness of change" (Spengler 71). Indeed, the chorus of the idea's pop-ups, listed above, describes our inability to be fully aware of something whose behavior falls outside the sphere within which we ascribe meaning. The second rendering of the earth-landing story makes clear that the aliens are us, the humans in the ecological drama we live in: Because what trees do—the immeasurable ways in which they have shaped the planet across millennia and made our own existence possible—happens too slowly for us to see, we do not register them as intelligent beings with agency. And while the alien story focuses on scales of time, some of the examples reveal that a similar effect arises because of discrepancies in scales of space: Entities are also incomprehensible when they are too massive.

An Avatar of Clonal Reproduction in Fiction

Authorial Motifs

Le Guin's and Powers's narratives explore both facets of continuity distinguished above—interconnection and persistence—illuminating why we may not instinctively conceptualize Pando as a single creature. This deeper understanding of how scale functions in tandem with iteration positions us to investigate the mapping of the latter. "Vaster Than Empires and More Slow" and *The Overstory* are *about* aspects of the science phenomenon we're discussing; the science manifests in them *literally*. To leverage the fi-sci framework and map the pattern of iteration, we need to identify a literary phenomenon that shares a shape with it, such that the science manifests in the fiction *figuratively*. Let's use our T expression to describe exactly what we're looking for:

a type of literary proliferation wherein multiplicity lies in iterations, each of which is a version of a singular entity—a proliferation that manifests similarly to a clonal organism.

I say "similarly" and not "in the same way as" because I seek an analog to a particular aspect of clonal reproduction, rather than one that matches each of its aspects in a one-to-one correspondence. In fi-sci analogies we make throughout this book there will be components of the source domain that map onto the target domain, and components that don't.[2] After all, analogies are similarities, not identicalities. In the present chapter, the feature of clonal reproduction I'm interested in for the purpose of exploring proliferation across fiction and science is *iteration*; a feature of clonal reproduction that I de-emphasize is *duplication*. Foregrounding iteration and backgrounding duplication helps capture the way I see the analogy between clonal reproduction and its parallel in fiction. And so, while pop culture and science fiction have ensured that "clonal" immediately brings to mind "exact copies," I am more concerned here with the "reproduction" component of clonal reproduction—specifically, iterative reproduction—and the location of multiplicity therein.

Iteration, then. "Iterations, each of which is a version of a singular entity" in literature pretty much defines "motif." We could look at motifs across literary works—tropes and archetypes. Or we could look at the leitmotif—the leading motif of a single work. To establish iteration on a grand, Pando scale, however, we will consider motifs that repeat across a single author's body of work. Why a single author's body of work? Because we seek an analog to the single root system from which spring all the clone ramets. Authorial motifs make a better pattern-mapping sandbox because they share a fundamental trait with the clonal organism form: A multiplicity of story segments/organisms that appear discrete are actually variations on one theme/genome. The F expression thus becomes,

In [authorial motifs] multiplicity lies in iterations, each of which is a {variation} on a singular <theme>.

The author who immediately comes to my mind in this configuration is Borges. The multiplicity here is not the number of Borges's individual texts of fiction, nonfiction, and poetry but rather the multiplicity of the same thematic iterations therein. This comes into stark relief in a graduate seminar on

2. See the discussion of metaphorical entailments and the invariance principle in the introduction.

Borges, for which students read a handful of his stories[3] back to back in the space of a semester. This seminar is not hypothetical; it took place at Florida International University in the fall of 2023. Week after week the students and I experienced a growing sense of déjà vu and began to wonder whether we were reading the same ideas over and over again, dressed in different costumes. This is not to say that Borges's work can be reduced to a simple outline; his are among the most complex pieces one will attempt to analyze. Rather, when we consider his stories as a collection—and when we read them in close succession—we find that variations on a theme emerge. Borges himself might have been sympathetic to this sentiment: He concludes "La esfera de Pascal" ["Pascal's Sphere"] musing, "Quizá la historia universal es la historia de la diversa entonación de algunas metáforas" ["Perhaps universal history is the history of the various intonations of a few metaphors"] (*Obras completas, 1923–1972* 638 [Weinberger 353]).

Avatar: The Borgesphere

We now make our way into the Borgesphere,[4] thus named because, as numerous scholars have discussed,[5] the same clutch of motifs recurs across Borges's oeuvre. The authorial motifs we will consider are more elaborate than the standard Borgesian trifecta described by Harold Bloom: "'Tlön, Uqbar, Orbis Tertius' begins with a disarming sentence (in Andrew Hurley's eloquent translation): 'I owe the discovery of Uqbar to the conjunction of a mirror and an encyclopedia.' That sentence is the purest Borges: add a labyrinth to a mirror and an encyclopedia, and you would have his world" (*How to Read* 58). "Mirror," "encyclopedia," and "labyrinth" are more like nodes in a Borgesian rhizome that "ceaselessly establishes connections" (Deleuze and Guattari 7). The authorial motifs on the other hand are intricate threads that are woven through and around the nodes in very specific patterns, ones so complex that perhaps we should call them microstories.

So what are these microstoried motifs? Once I began collecting them from Borges's fiction a neat recursion came to light: The most prominent and repeated iterations are actually *about* iteration. I found that each of the many examples I located could be filed under one of four overarching types, and that these types are arranged along a spectrum of iteration: At one end is an

3. I use *stories* as an umbrella term for Borges's texts, which includes texts presented as fiction and as nonfiction.

4. I owe the discovery of the term *Borgesphere* to the conjunction of Nicholas R. Cabezas and a WhatsApp group.

5. See, for example, Christ; Alazraki, "Borges' Modernism"; Aizenberg; Fishburn.

absolute absence of iteration, and at the other end some kind of organizational scheme is called upon to tame the chaos.

I. Singular (ideal, unity): One entity is all entities
II. (De)Gradation: Each successive version moves further from the ideal
III. Variance: Entities differ depending on their contexts and arrangements
IV. Systematization: Differentiation of entities is formalized in a classification system

Performing my own version of type IV, I list the formulations that belong to each type and, in parentheses, the number of Borges stories in which I've located examples of the given formulations (the appendix lists the titles of these stories and reproduces examples).

I. Singular

- (Given eternity) One object is all objects; one man is all men; all men are one man (9)
- One object or symbol encompasses everything, or stands for God, or the word of God, or contains the totality of all things (6)
- One object, symbol, or entity has many random guises but is in fact one and the same in essence (2)

II. (De)Gradation

- (De)Gradation of ideal forms; categories; representations (3)
- A 1:1 map-territory relation (5)
- A suggestion or explicit mention of infinite recursion or regression (8)
- Doubt about an entity's—or one's own—ontological status; fiction overtaking reality; the imagined imposed onto reality (4)
- Anachronistic precursors (2)

III. Variance

- Permutation (8)

- The same words are different depending on who expresses them (2)
- Entities are different at different moments (2)

IV. Systematization

- Classification and taxonomy; the invention of a system for such (4)
- Strange language; the creation or existence of a (universal) language (6)
- Inventorying an author's works (2)

The above classification is not scientific; other readers will parse motifs and formulations differently. It is also by no means exhaustive; many are the Borges narratives I did not sift through in search of motifs. I do hope, however, that the above conveys how often these particular authorial motif ramets pop up, and how fitting it is to characterize at least a considerable portion of Borges's oeuvre as variations on a theme. As observed earlier, the four types (singular; (de)gradation; variance; systematization) lie on a spectrum of iteration: from an absolute lack of iteration to a systematization of iterations. Iteration, then, can be considered the authorial megamotif of the Borgesphere.

The megamotif of iteration arises again and again in stories Borges wrote at different times, and which appeared in different publications. Walking among stems of the Borgesphere forest, readers may perceive each trunk as a distinct microstory. And while this is of course technically true, so many of them share a genome: They are variations on the theme of iteration.

Writing as Rewriting

Borges concludes his story "Pascal's Sphere" with a proposal I quoted a few paragraphs back: "Perhaps universal history is the history of the various intonations of a few metaphors." Jaime Alazraki traces the idea behind this proposal to an article Borges published in *Sur,* the penultimate sentence of which is "La literatura es fundamentalmente un hecho sintáctico" ["Literature is fundamentally a syntactic fact"] ("Elementos" 161 ["Borges' Modernism" 168]). Alazraki explains that Borges ascribes to "a theory of literature which views writing as rewriting" (168)[6] and that this can be "regarded as the cornerstone

6. "Syntactic" is the adjective form of "syntax," which comes from the Greek *syn-* ("together") and *tassein* ("arrange"), so a "systematic arrangement of parts" ("Syntax"). When qualifying "literature," "syntactic" conveys that literature is a matter of arrangement and rear-

of his poetics" (169). Evelyn Fishburn recognizes the same when she mentions Borges's "delight in the reworkings of a few basic ideas" (1); Bloom does, too, noting, "Borges came to see all great literature as a single cyclic, storytelling poem, written and rewritten throughout the centuries" (*Genius* 682). Edna Aizenberg writes of Borges's "insistence that literature is a *syntactical* event in which images are not created but *reelaborated*" (7).

Saying that literature is syntactical, that writing is rewriting, is saying that literature's modus operandi is variation. That is, in the formulation of literature as syntactical, the emphasis is not so much on new, unique outcomes (analogous to offspring of gametic reproduction) as it is on iterations of existing outcomes (analogous to the phenotypes of offspring of clonal reproduction). And Borges did not just *see* literature that way, he also *performed* literature that way, in the sense of practicing what he preached. As Alazraki notes, while the views of Borges resonated with those of the Russian Formalists, Borges "did what the Formalists did not do: he incorporated some of his ideas about literature into his creative writing, thus amalgamating theory and craft" (171). And indeed, we have seen this in our analysis of Borges's authorial motifs: His megamotif is iteration. Thus Borges both iterates (perpetually alluding to and drawing from precursors) and very often writes about *others* iterating. Iteration is his paramount philosophy, mode, and subject.

Though the preceding analysis focuses on charting Borges's authorial motifs and demonstrates that iteration looms large in his corpus, the iteration lens is not the only one through which we can examine his stories. The convergence lens we applied to the reading of *Adios, Happy Homeland!* in chapter 2 would reveal copious material, as evidenced by volumes devoted to thoroughly tracing allusions and references in Borges's work.[7] Indeed, the Borgesphere is tremendously transtextual. Daniel Balderston remarks that "fiction as patchwork, or as what Nahum Cordovero in 'El inmortal' termed the 'coat of many colors,'" is "one of the fundamental features of Borges's art" (xix). Rather than prescribing a certain lens to a given narrative, the duet proliferation chapters seek to show what can be illuminated when we apply one or the other perspective to a body of literature—the kind of pattern that each application allows us to map.

---

rangement, in other words, the placing of what is already extant in a different order: writing as rewriting.

7. See, for instance, Fishburn and Hughes; Balderston; Berveiller.

TABLE 9. Proliferation (iterations) flowchart

	Phenomenon	Example	Expression
Science	1 Clonal reproduction	2 When aphids reproduce, each offspring sets out as a genetic replica of the progenitor and is identical to its siblings.	3 In [clonal reproduction] multiplicity lies in iterations, each of which is a {copy} of a singular <progenitor>.
Template	—	—	4 In [a certain type of proliferation] multiplicity lies in iterations, each of which is a {version} of a singular <entity>.
Fiction	7 Authorial motifs	6 Works by Jorge Luis Borges	5 In [authorial motifs] multiplicity lies in iterations, each of which is a {variation} on a singular <theme>.

Mapping Entailments of Proliferation: Iterations

Variations

Something I touched on earlier now emerges as a metaphorical entailment. Like Pando's ramets across its forest, Borgesian motifs pop up throughout the Borgesphere; multiplicity in both cases is located in iteration. What the two have in common is that both are iterations of a single entity. The metaphorical entailment is that both the ramets and the authorial motifs manifest variations. Each Borgesian motif—though an iteration of one theme—lives in a different context, which leads to a different expression of the central idea. Whether it is because of the setting or tone of each story, the period during which the story was written, the way several motifs interact intratextually, or other factors, the theme will transmute; its manifestations will not be identical.

Analogously, Pando's ramets are versions of the parent: Though they contain identical genetic markers, random mutations ensure that there will be diversity among them. Epigenetics adds to variation among clones. Pando spans 106 acres, so it is conceivable that some of its stems will sprout on a hill while others will be closer to the water. The tree's 47,000 trunks serve as homes to thousands of species, each of whose individual habitat affects a single stem in a slightly different way. In both fiction and science, iterations (on a theme; of a genome) emerge from a singular source and have varied expressions. Table 9 presents the flowchart for the Authorial Motifs || Clonal Reproduction fi-sci analogy.

Duet

The more general Proliferation || Reproduction fi-sci analogy—that encompasses chapter 2's Transtextuality || Gametic Reproduction fi-sci analogy and this chapter's Authorial Motifs || Clonal Reproduction fi-sci analogy—plays better as a duet: Transtextuality is to authorial motifs as gametic reproduction is to clonal reproduction. The counterphenomena in fiction support the conceptualization of the respective counterphenomena in science. The processes of gametic and clonal reproduction are indiscernible to the naked human eye; aspects of some products of these processes—like Pando's forest being one organism—strain conceptualization. Viewing the former as analogous to transtextuality in concert with viewing the latter as analogous to authorial motifs renders more robust our overall understanding of biological reproduction.

This brings us back to why, among the characteristics of reproduction, I have singled out multiplicity. Multiplicity enables the analogy of proportionality[8] that explains why chapters 2 and 3 form a duet. Analogies of proportionality take the form A:B :: C:D, whereby A is to B as C is to D. The analogy of proportionality that emerges when we consider these two chapters together, visually represented in figure 3, is *gametic reproduction* ([A]) *is to transtextuality* ([B]) *as clonal reproduction* ([C]) *is to authorial motifs* ([D]). Multiplicity connects the pattern common to [A] and [B]—which is convergence—to the pattern common to [C] and [D]—which is iterations—because it is in convergence (fi-sci analogy I) and iterations (fi-sci analogy II) that multiplicity is located. In showing the interconnections of these relations, the diagram in figure 3 makes the case for reading chapters 2 and 3 as a duet: The overlaps are simultaneously vertical (convergence, iterations) and horizontal (proliferation in organisms, proliferation in texts).

The benefit of focusing on the multiplicity characteristic is that it allows us to explore how [A]:[B] is proportionally analogous to [B]:[C]. And the benefit of *that* is that we gain a new insight into the maneuvers and devices

8. This particular kind of analogy was first proposed by Tommaso de Vio Cajetan in his *De Nominum Analogia* [*On the Analogy of Names*] (1506): "The name *proportionality* is given to a similitude of two proportions; e.g. we say that eight is to four as six is to three, because both are twice as much in proportion, etc." (25).

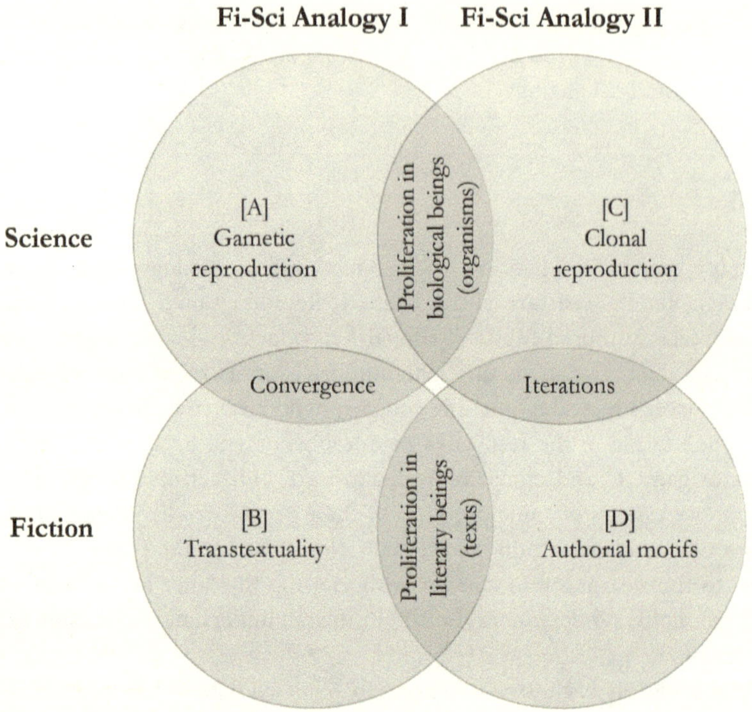

FIGURE 3. Proliferation duet.

of fiction. What we notice now that previously—in the absence of fi-sci pattern mapping—we likely had no occasion to discern is a relationship between transtextuality and authorial motifs: They are both mechanisms by which literary beings (texts) proliferate; they contrast in how multiplicity manifests in each. Proliferation through transtextuality takes the pattern of convergence: Other texts proliferate by converging in the new text and living on. Convergence implies multiplicity (a singular thing cannot converge with itself). Proliferation through authorial motifs takes the pattern of iterations: The author's texts proliferate with every iteration of the motifs. Iteration implies multiplicity (that something is iterated means there now exist more than one of it). We wouldn't have gotten here without connecting these fi to the sci of reproduction, and, further, without focusing on the characteristic of multiplicity. It is by viewing transtextuality and authorial motifs through the lens of multiplicity that we can detect this relationship between them.

PART 2

NONCLASSICAL PHYSICS

CHAPTER 4

Warp

This chapter is our first foray into case studies involving nonclassical physics. When differentiating perceptual groups in the introduction, I noted that interactions of objects at the cosmic scale are described by relativistic mechanics. The theory of relativity accounts for features of space and time that can be difficult to visualize. To our minds space is governed by Euclidean geometry, which assumes a flat surface, and wherein axioms such as "two parallel lines cannot meet" bear out in our day-to-day observations. Geometry at the cosmic scale, however, is less than intuitive because it is non-Euclidean; a curved surface entails altogether different axioms, which describe phenomena that do not always coincide with lived experiences.

How might we wrap our minds around the strangeness of non-Euclidean space? In this chapter we will attempt to do so by seeking a maneuver in fiction whose pattern can be mapped onto an extreme instance of spacetime curvature: the wormhole.

Spacetime Curvature

Geometry as codified by Euclid involves axioms and theorems about entities such as lines, triangles, and circles. An example of a Euclidean theorem: The sum of a triangle's angles is 180 degrees. Euclidean geometry was considered self-evident: "For well over two thousand years, people had believed that only one geometry was possible, and they had accepted the idea that this geometry described reality" (Banchoff). But as cosmologist Paul Davies notes, while "the theorems work correctly on blackboards and sheets of paper . . . they do not work on curved or warped surfaces." He adds that "pilots and navigators are well aware of this, and they have to use different geometrical rules to cope with the Earth's curvature." So while Euclidean axioms and theorems are "watertight . . . they come with one important proviso: they apply to *flat* surfaces" (34). Accordingly, different geometrical rules are relevant when we consider, for instance, an atlas, versus when we consider a globe.

So far we have been talking about the three dimensions of space, but in the next chapter of geometry's story another dimension joins the crew: time. Einstein's theory of special relativity (1905) showed that "space is linked to time in a manner that makes it natural to consider the whole package—*spacetime*—together" (Davies 37). Three years later mathematician Hermann Minkowski "worked out how to modify the rules of Euclidean geometry to describe four-dimensional spacetime" (Davies 37). In his 1915 general theory of relativity, Einstein introduced a twist, so to speak: Spacetime can warp. The mass of an object (e.g., a star) curves the "fabric" of spacetime around it, such that space and time "behave" according to different rules altogether. For instance, spacetime curvature can cause two initially parallel lines to intersect—an outcome that directly contradicts the Euclidean postulate that defines parallel lines as lines that cannot meet. So why do we not see parallel lines meeting in our everyday experience of reality? At the human scale the force of gravity is too weak to make a noticeable difference to the flows of time or space, and this keeps the two types of dimension separate in our minds. When making calculations at the cosmic scale, however, one needs to take into account all four dimensions: three of space, and one of time.

One extreme instantiation of spacetime warping is a wormhole. Wormholes are hypothetical structures for which "there is no experimental evidence" ("Wormhole"). Despite their hypothetical nature, we can talk about wormholes as science rather than "just" science fiction because they arise as theoretical, mathematical solutions of Einstein's equation for the general theory of relativity (Woo). Wormholes are famous for their (hypothetical) capacity to connect regions of spacetime that are causally disconnected. Regions that are causally disconnected are ones that cannot bear upon each other. "Because the speed of light, acting as a universal speed limit, is finite, regions of the Universe which are well separated are unable to communicate with each other. They are said to be causally disconnected" ("Causality"). These regions are so far apart that in an expanding universe light from one cannot get to the other. It follows that an event occurring in one region cannot affect or be affected by an event occurring in the other. To picture why this is the case, let's think in terms of light cones.

Physicists use the concept of light cones to divide spacetime into sections "relevant to particular observers" ("Light Cone"). Imagine a point in spacetime that represents a particular entity, *e*. The light cone "behind" *e*—of which *e* forms the apex—encompasses all possible entities that could have led to, caused, had an impact on *e*. The light cone "in front" of *e*—that is, stemming from *e*—encompasses all possible entities that *e* might lead to, cause, impact. Regions outside these two cones are causally disconnected from *e*. "Since we ourselves can never exceed the speed of light, whatever we do we are always compelled to stay within our future light cone as we move into the future. Likewise, we must have reached our present point from some location in our past light cone" ("Light Cone"). Effectively, this means that a person at spacetime point *e* cannot cross paths with a person who exists outside the two light cones. Unless, that is, there's a wormhole connecting these two points. A wormhole can warp spacetime in such a way that causally disconnected points can be connected. Let's describe wormholes with the S expression,

> [Spacetime] curves such that {regions} on either side of an effectively uncrossable boundary can become connected.

We refine the S expression into a generalized version; to render the T expression, we replace the science particulars with discipline-neutral terms but keep the rest as is to maintain the pattern:

> [A framework] curves such that {entities} on either side of an effectively uncrossable boundary can become connected.

An Avatar of Spacetime Curvature in Fiction

How does our T expression manifest in fiction? Otherwise put, what kinds of stories follow this shape, such that we could substitute the items in brackets with the corresponding items in the realm of fiction?

Metalepsis

Our search is for a story where the framework curves in a way that allows the meeting of two entities that, owing to their "causally disconnected" regions—their existence on either side of an effectively uncrossable boundary—should not be able to meet. In fiction, the framework is the narrative milieu, or "storyfabric"; the entities are the characters; the characters' regions are their ontological strata. The wormhole that curves the storyfabric such that characters' different ontological strata can intersect is none other than metalepsis, "literally 'a jump across'" narrative boundaries (Kukkonen 1).

Narratologists generally distinguish between two varieties of metalepsis: rhetorical and ontological. We find rhetorical metalepsis in Kurt Vonnegut's *Slaughterhouse-Five* (1969), when the narrator talks to the reader, and makes it clear that characters such as Billy Pilgrim are his own creations. Ontological metalepsis, on the other hand, involves the permeation of narrative strata, as in Michael Ende's *Die unendliche Geschichte* [*The Neverending Story*] (1979). Here, protagonist Bastian reads a book that is also titled *The Neverending Story.* In this embedded book, a character likewise reads a book titled *The Neverending Story,* which turns out to be the very book Bastian is reading. Bastian then enters the world of the story, interacting with the characters therein and becoming part of their story. The protagonist thus traverses ontologically separate narrative strata. Marie-Laure Ryan explains the distinction as follows:

> Rhetorical metalepsis opens a small window that allows a quick glance across levels, but the window closes after a few sentences, and the operation ends up reasserting the existence of boundaries. This temporary breach of illusion does not threaten the basic structure of the narrative universe. . . .
>
> Whereas rhetorical metalepsis maintains the levels . . . distinct from each other, ontological metalepsis opens a passage between levels that results in their interpenetration, or mutual contamination. . . . In a narrative work, ontological levels will become entangled when an existent belongs to two or more levels at the same time, or when an existent migrates from one level to the next, causing two separate environments to blend. (207)

The existents we are presently considering are characters who do not just glance across strata, but ones whose ontological strata fully intersect. Our focus is thus on ontological metalepsis.

If ontological metalepsis is what we seek, we should be able to designate Flann O'Brien's *At Swim-two-birds* (1939) as our fiction example. The novel features numerous jumps across ontological levels, as for instance when fictional characters conspire against their different-stratum-fictional author. However, once we take into account the perspective and therefore experience of the reader vis-à-vis the metalepses in this novel, we find that the kind of ontological metalepsis at play in *At Swim* does not align with the characteristic of wormholes that is the focus of this chapter, which is that they are deeply counterintuitive. In what follows I explain why.

Earlier, when picturing the light cones that spread out in either direction of *e*, you, this chapter's reader, took on the role of an external observer. As someone located outside the system in question, someone who could see at once regions that pre-wormhole were unconnectable, you do not find their connection to be a challenge to your perception. For an individual situated inside the light cone, however, access to a point outside the light cone is counterintuitive. Analogously in fiction, the reader who maintains their perspective as one outside the story's system may find it quirky when characters who are on different ontological strata interact, but the reader is not likely to find this shocking. After all, this seemingly impossible communication is being experienced by others, others who are moreover fictional. In the case of *At Swim*, the reader is led to maintain the role of outside observer throughout their reading of the novel. When presented with ontological metalepsis, the reader does not find themselves at the mouth of the wormhole, with sudden access to characters with whom they did not previously expect to be able to interact; rather, this counterintuitive connection is happening to someone else. For example, the reader learns that fictional author "[Dermot] Trellis has absolute control over his minions [characters] but this control is abandoned when he falls asleep" (47–48). Hence, upon drugging him, his characters are able to do as they please rather than be forced to conduct themselves according to his writing. But since the reader recognizes that Trellis is himself a character, this connection is happening among the set of O'Brien's characters, and the reader watches it unfold, so to speak, from above. Let's characterize this, no pun intended with respect to the novel's title, as bird's-eye-view metalepsis, or bird metalepsis for short.

What we are looking for, however, is someone who experiences the full shock of metalepsis—a reader whose counterpart in our spacetime system is bizarrely granted access to a region heretofore inaccessible. We seek a story that ushers its reader to align or identify with a character that finds themselves in counterintuitive contact with another character they should not be able to meet, to experience the dizzying twist of metalepsis such that the reader shares—even empathizes with—the character's shock. Let's refer to this second type—again no pun intended, this time with respect to wormholes—as worm's-eye[1]-view[2] metalepsis, or worm metalepsis for short.

There is another characteristic of our sought-after text that becomes apparent when we consider where exactly a change is effectuated. In the Greek μετάληψις (metalipsis) from which "metalepsis" derives, the noun λῆψις (lipsis) in turn derives from λαμβάνω (lamvano), which means "take hold of," and the prefix μετά (meta) denotes "beyond," or "transcendence." In his account of the term Genette notes that he is using "metalepsis" "with this specific sense: 'taking hold of (telling) by changing level'" (*Narrative Discourse* 235n51); for the purposes of studying narrative he interprets "take hold of" as "tell," and "transcendence" as "changing level." For Genette, changing levels means transgressing a boundary, "a shifting but sacred frontier between two worlds, the world in which one tells, the world of which one tells" (236). Recall that this is "literally 'a jump across,'" whereby "the boundaries of a fictional world are glanced, travelled or transported across" (Kukkonen 1). The entity transgressing this boundary is variously the narrator, narratee, or a character (Genette 234)—let's use Ryan's terminology and collectively refer to these entities as existents.

And so, in short, metalepsis occurs when an existent traverses a narrative boundary. This conclusion suggests that the author's act (application) of metalepsis bears out in the acts of the existent—the one who does the traversing. I wish to extend the scope of metalepsis such that it pertains also to shifts *in the milieu* within which existents operate. In this expanded purview, the storyfabric does not stay still, unaffected, while characters jump across boundaries; rather, the author's use of metalepsis effects a change in the storyfabric, and the characters—those "on" the fabric, who travel in its medium—are moved along with it, as though riding a wave. Otherwise stated, what the author's use of metalepsis affects are the fabric's contours. Here characters are not jumping; like an individual who is about to enter the mouth of a hypothetically traversable wormhole, they are continuing (foreshadowing: "Continuity!") along the same ontological stratum on which they had set out. It is the ground beneath

1. Rachel Hoskins of the Woodland Trust notes that worms do "not really" have eyes. "Instead, they have cells called receptors that can sense whether it's light or dark." Per Purschke et al., "Most annelids respond to light and possess some type of light-sensitive structure, which may comprise different types of PRCs, ciliary or rhabdomeric PRCs. However, only light-sensitive structures equipped with shading pigments, the structural prerequisite for detecting the direction of light, are generally called eyes, irrespective of whether they are capable of vision." It is in this sense that I continue to refer to worms' eyes in this chapter.

2. The fact that "worm's-eye view" has previously been established as the obverse of "bird's-eye view" (see "Point of View in Photography") is a kind of metaphorical entailment.

them that shifts, altering the path of the characters along with it. Thus, change is effectuated *in the framework,* not *by the characters.*

Synontology Detour

Before we continue the search for a story whose metalepsis more closely aligns with the counterintuitive aspect of wormholes, let's carefully examine the matter of bird metalepsis. Such a detour is worthwhile because establishing the particulars of bird metalepsis positions us to set that reader experience in contrast to the one we are ultimately seeking: that of a reader who is being shepherded to follow characters' perspectives on the ground.

The detour begins in Riverside, Iowa, where visitors are able to participate in a peculiar kind of commemoration: a celebration of the future birth of *Star Trek*'s Captain James T. Kirk. Travelers to Reichenbach Falls, Switzerland, likewise pay homage to a fictional event, marked by a plaque that, in three languages, memorializes the fateful altercation between Arthur Conan Doyle's Sherlock Holmes and James Moriarty. This treatment of fictional entities as though they were nonfictional is not limited to commemoration. You may purchase a Wonka Bar at a convenience store, publish a scholarly article in *HolQed,* the peer-reviewed Klingon journal, converse in the elvish language Sindarin, and play college muggle Quidditch. These are just a few instances of a quirky phenomenon whereby readers interact with fictional entities outside the narratives in which these entities originate.

I designate such interactions synontological events[3]—using the Greek prefix "syn" to denote "together" or "with"—as these events occur at the confluence of two ontologies: that of fiction, and that of nonfiction. For instance, the synontological event of purchasing Wonka candy conflates the fictional world of Roald Dahl's *Charlie and the Chocolate Factory* and the nonfictional world in which Nestlé's Willy Wonka products such as Everlasting Gobstoppers and Wonka Bars can be bought at the store. In purchasing Wonka candy, one is interacting with a heretofore fictional entity in the nonfictional world, thereby conceptually accessing two ontological strata. Like metalepsis, synontology describes a shift from one stratum to another—a shift that turns on the audience's playful interaction. This play-along involves different types of mental maneuvers; on our detour we will investigate synontology via rhetoric, metalepsis, and cognition in turn.

3. See Trauvitch, "Synontological Spaces."

Rhetorical Intersubjectivity

We commence with the "special kind of rhetorical intersubjectivity"[4] that enables synontology. In "Truth in Fiction: A Reexamination of Audiences," Peter Rabinowitz holds that the "duality" inherent in reading fiction—the fact that "events must be treated as both 'true' and 'untrue' at the same time"—gives rise to four audiences (125). These are the actual audience, the authorial audience, the narrative audience, and the ideal narrative audience; Rabinowitz sees these as "different levels on which a reader operates simultaneously" (130n18). For my purposes, the first three of these audiences are the most important.[5] Phelan notes that typically, in the narrative-audience role, readers treat the characters and events as real, whereas in the authorial-audience role, readers "operate[ ] with the tacit knowledge that the characters and events are synthetic constructs rather than real people and historical happenings" (*Narrative as Rhetoric* 215). The awareness of *both* is the duality, or double consciousness of rhetorical reading.

At the first level of consciousness readers take the synthetic constructs as real, and simultaneously at the second level of consciousness readers retain the awareness that the synthetic constructs are invented. I propose that in synontology we read with an unusual, triple consciousness: In addition to the typical processes of (1) taking what is fictional as real and (2) retaining the awareness that it is fictional—an awareness of both the narrative and the authorial-audience modes—readers take an additional step (3) of pretending that what they know is fictional is real. Phelan writes, "This pretending is a third step because it comes after (in a conceptual rather than a temporal sense) taking on the roles of authorial and narrative audiences. In other words, readers have the double consciousness but then playfully pretend that only the single consciousness of the narrative audience is relevant."[6]

Notably, while the three levels are experienced concurrently, they build on each other. There is no level 2 (the retained awareness that the synthetic constructs are invented) without level 1 since level 1 (taking the synthetic constructs as real) is the awareness that level 2 retains. Likewise, we cannot do away with level 2 (the retained awareness that the synthetic constructs are invented) because then there would not be any awareness to playfully deny. Playful denial is the core characteristic of synontology.

---

4. James Phelan, personal communication with the author, April 2017.

5. See Sarah Copland and James Phelan, "The Ideal Narratee and the Rhetorical Model of Audiences," regarding a revised model of the ideal narrative audience.

6. James Phalen, personal communication with the author, August 2024.

Dizzying Metalepsis

Our focus here is on synontology that exists outside the borders of a given narrative. In other words, we are examining synontology that occurs beyond the readers' experiences of a text. Your purchase of a Wonka Bar, for instance, takes place outside your experience of reading Dahl's *Charlie and the Chocolate Factory*, since your purchase is not described in the novel. A visit to Riverside's monument takes place outside the experience of watching *Star Trek*, as this visit is not featured in any *Star Trek* episode or movie.

Synontology regularly occurs within narratives as well—through the process of metalepsis, since metalepsis involves the meeting of entities from two different ontological levels of a narrative. However, metalepsis provides us with a springboard rather than a framework for understanding synontology that occurs beyond the text. Metalepsis is necessarily a text-bound device, which involves a paradoxical situation that cannot occur in actuality. *At Swim* is a perfect example of this kind of situation. Moreover, fictional entities do not literally jump across into actuality; what we interact with when we purchase a Wonka Bar or play Quidditch is a representation, copy, or re-enactment of a fictional entity. If one were to repaint in magenta a Kwik-E-Mart that exists in actuality (and several did physically exist as part of a marketing campaign in 2007), one would not whatsoever affect the walls of its fictional version, which continue to be orange in *The Simpsons*.[7] Thus, while metalepsis and synontological spaces in actuality both pertain to the convergence or crossing of ontological states, they describe different negotiations of these states—one within the text and one beyond it.

Though different phenomena, both synontology and metalepsis exude the riveting scent of paradox. Synontology is not paradoxical per se, and it does not involve any literal transgression, because nothing impossible takes place. However, the brain of a Riverside tourist is doing interesting things as they negotiate two ontological states. What happens in our minds that allows us to commemorate the occasion of a fictional birth—let alone one that is set in the future of our earthly, nonfictional timeline? With what mindset, and with which cognitive tools do we interact with fictional entities beyond our experience of reading about them or watching them on TV? That we do so easily does not diminish the quirkiness of the situation; that we do so often indicates that there is something rewarding about this interaction. There is a fascinating cognitive puzzle here.

---

7. Stephan Packard, personal communication with the author, November 2015.

Cognitive Foundation

Ontological metalepses are those that engender truly paradoxical situations. They suggest the incomprehensible because in addition to being untrue, they are also impossible. Fictional characters cannot conspire against their author. And yet, in *At Swim* they do, and we are somehow able to envisage this situation. If incomprehensible phenomena defy ontological understanding, why is it that we are able to seamlessly accept them in fiction (to the extent that we can imagine them)? How exactly does this acceptance work?

The cognitive substructures of synontology are flexibility and dissonance. Cognitive flexibility explains how readers access multiple ontological states, and cognitive dissonance suggests that perhaps the tension evoked by the whimsical juxtaposition of the fictional and the nonfictional is experienced positively. In their discussion of learning and knowledge acquisition, Spiro and Jehng define cognitive flexibility as "the ability to spontaneously restructure one's knowledge, in many ways, in adaptive response to radically changing situational demands" (165). Cognitive flexibility is at play in synontology, as one must almost simultaneously hold two ontological perspectives. Consider a fan's visit to the Riverside monument. They must indeed display something comparable to "the ability to spontaneously restructure [their] knowledge . . . in adaptive response to radically changing situational demands." The situational demands are the demands made by fiction and the demands made by actuality. The demands do not so much rapidly change as they are merged and accessed simultaneously, but they are decidedly radically different. A reader approaches fiction with a mindset different from that with which they approach nonfiction (see the introduction). For instance, when reading fiction one is typically ready to suspend one's disbelief. One is more willing to accept outrageous concepts and flex one's imagination. This is, crucially, the reason we are able to accept in fiction that which is inconceivable in nonfiction, and thereby ease our understanding of the similarly patterned, difficult to conceptualize thing in science.

Our fan accesses two ontological states—that of perceiving the actual, nonfictional monument, and that of experiencing the fictional world of Captain Kirk—and they do so via cognitive flexibility. The fan is able to concurrently inhabit both realms in touring a synontological space by rapidly switching mindsets or accessing both at the same time. Elen et al. describe cognitive flexibility as

> the disposition to consider diverse context-specific information elements while deciding how to solve a problem or to execute a (learning) task in a

> variety of domains and to adapt one's problem solving or task execution in case the context changes or new information becomes present. . . .
>
> Given that context-specific information elements are taken into account, being cognitive [*sic*] flexible implies that one considers both the context and the information at hand. (2)

Though the authors are discussing cognitive flexibility in reference to learning, problem solving, and task execution, we can apply their delineation to our analysis. In terms of synontology, then, what is the information at hand, and what is the context? The information is a given entity's ontological status. The fan possesses information that Captain Kirk is fictional, and that the monument commemorating Captain Kirk's future birth is nonfictional—an actual, engraved stone that exists in an actual town. The context, on the other hand, is the synontological space itself, wherein the two ontological states are merged. The fan operates within this context when they visit the monument.

Within the context of synontology, our information about one state must be reconciled with our information about the other state, and we appreciate the seemingly paradoxical juxtaposition. Essentially, the fictional must be conciliated with the nonfictional such that the two make sense contemporaneously and colocationally. Applying the definition provided by Elen et al. to our investigation, we can determine that "being cognitive [*sic*] flexible implies that one considers both" the synontological space ("context") and a given entity's ontological status ("information") "at hand." Tying this to Spiro and Jehng's definition and repositioning it for our purposes, we can further establish that cognitive flexibility "is the ability to spontaneously" reconcile pieces of information ("restructure one's knowledge"), "in many ways, in adaptive response to" seemingly incongruous ("radically changing") juxtapositions ("situational demands"). Those who participate in synontology, it would appear, are quite cognitively flexible.

So far we have noted the abilities of synontology participants, but what of their motivations? What does one gain from visiting an actual place inhabited—in fiction—by a fictional character, or making a pilgrimage to commemorate a future, fictional event? I suggest that we interact with fictional entities in our nonfictional reality because we seek novelty and are beguiled by the peculiar confluence of fiction and nonfiction. I suspect a somewhat similar sentiment is brewing when we encounter oxymorons, and perhaps even synaesthesia. We delight in seeing how two things that supposedly should not work together *do.*

The fan visiting the Riverside monument enjoys the jarring sensation of paradox, and this is where cognitive dissonance comes into play. Leon

Festinger developed cognitive dissonance theory and based it on the premise that "the individual strives toward consistency within himself" (1). Supplanting "inconsistency" with "dissonance" and "consistency" with "consonance," Festinger described cognitive dissonance theory with two hypotheses:

1. The existence of dissonance, being psychologically uncomfortable, will motivate the person to try to reduce the dissonance and achieve consonance.
2. When dissonance is present, in addition to trying to reduce it, the person will actively avoid situations and information which would likely increase the dissonance. (3)

Joel Cooper writes that Festinger "made a very basic observation about the lives of human beings: we do not like inconsistency. It upsets us and it drives us to action to reduce our inconsistency. The greater the inconsistency we face, the more agitated we will be and the more motivated we will be to reduce it" (2). Cooper notes that "a pair of cognitions is inconsistent if one cognition follows from the obverse (opposite) of the other" (6). Thus, "if a person holds cognitions A and B such that A follows from the opposite of B, then A and B are dissonant" (6).

I characterize what readers of synontology experience as cognitive dissonance because they indeed hold "cognitions A and B such that A follows from the opposite of B," where "opposite" is broadly interpreted. For example,

- Let cognition A = that fact that you are visiting the site where Captain Kirk is to be born, and commemorating this future event;
- Let cognition B = the knowledge that Captain Kirk is a fictional character.

Cognition A follows from the converse of cognition B in that you would typically visit a site where something *non*fictional happened and commemorate an *actual* event (one that moreover occurred in the past). This is at odds with cognition B, according to which the event, like the character, is fictional. Given this formulation, any interaction with a fictional entity that takes place in our nonfictional world gives rise to cognitive dissonance. Of course, synontology does not usually produce the discomfort or vexation that Festinger and Cooper describe; even an accidental visitor to Google Street View's TARDIS—the Time And Relative Dimension(s) In Space craft that traverses space and time from the series *Doctor Who*—right there in plain digital view at Earl's Court Road, London—is unlikely to feel distressed about this find. However, rather than indicating that synontology does not involve cognitive dissonance, this actually suggests that cognitive dissonance—and inconsistency—is not always infelicitous. I am proposing that the contemporaneous and collocational confluence of ontological states gives rise to an inconsistency that is experienced *positively.* Adds Phelan, "The audience in a synontological situation is aware of the invention and the use of imagination at the fictional level; the playful denial builds on that awareness by extending the imaginative act."[8]

Indeed, instead of striving to eliminate the inconsistency—their cognitive dissonance—visitors of synontological spaces actively seek it out. The tension produced by the curious juxtaposition of fiction and nonfiction is experienced favorably. The inconsistency is what makes it fun! Part—perhaps most—of the thrill of shopping at Kwik-E-Mart emerges from the ability to purchase heretofore fictional products—something that, like suddenly finding yourself in *The Simpsons*' Springfield, should not be possible. Instead of being avoided, this cognitive dissonance is sought, and cognitive flexibility sustains the delightful feeling of dissonance. Happy synontology participants are ones who have the flexibility to experience this dissonance.

"Continuidad de los parques"

The examples of synontology are different from those of metalepsis because the former are unbound from the text. But synontology does have something significant in common with bird metalepsis: Drawing on the rhetorical model of audiences, we can say that both subordinate the narrative audience to the authorial audience. In synontology and in elevated, all-seeing bird metalepsis like that in *At Swim,* the readers are on the outside. It is this reader position that precludes both synontology and bird metalepsis from suitability to this chapter's fi-sci analogy. Pre-detour we established that what we require is a story that routes the reader to experience metalepsis from the perspective of the existent who is implicated therein. The reader should be shocked by the strange access to an entity that should be inaccessible. Rather than viewing the situation from above and seeing the entire layout, the reader we seek should be prodded along through the narrative's weird tunnel. A narrative that takes readers down just such a bewildering wormhole is Julio Cortázar's "Continuidad de los parques" ["Continuity of Parks"] (1964). This story's ontological metalepsis arises when the text suggests that a man reading a book is about to be killed by a character featured in said book. I write "suggests" because Cortázar never actually says this happens. What I would like to investigate

8. James Phelan, personal communication with the author, August 2024.

here, then, are the narrative devices that shepherd this story's reader to reach a conclusion that is impossible.[9]

Frederick Luis Aldama notes, "Already with the title we're . . . being *guided* to create a continuity between [the] two spaces" of the man reading and the plot of the book he's reading. "So, already . . . Cortázar is *directing*—he's kind of loosely setting up—that continuity that we're going to create in our minds" (16:25–17:00, emphasis added). Another maneuver that guides the reader to deduce an ontologically metaleptic outcome (one that is not actually explicitly stated by Cortázar) is carefully placed repetition. Initially, we perceive the aforementioned two spaces as existing on different ontological strata: One is the stratum of the man reading the novel; the other is the stratum of that novel's plotting couple. But as we make our way through the story we come across more and more echoed concepts and words, which reach a critical mass that leads us to merge the two strata.

Stratum [I]'s "la tranquilidad del estudio que miraba hacia el parque de los robles" ["the tranquility of his study which looked out upon the park with its oaks"] (Cortázar 9 [Blackburn 55]) and "bajo los robles" ["under the oak trees in the park"] (9 [55]) reappear on stratum [II] as "los árboles y los setos" ["the trees and hedges"] (10 [56]) and "la alameda" ["the avenue of trees"] (10 [56]). These echoes on stratum [II] won't necessarily raise a flag; after all, parks, oaks, and avenues are common enough. Then stratum [I]'s "el mayordomo" ["manager of his estate"] (9 [55]) pops up as stratum [II]'s "El mayordomo" ["The estate manager"] (10 [56]) so this, being less common, might perk up some reading ears. Still, it is probably a coincidence, right? Perhaps, we reason, estate managers are regular features of Cortázar's stories.

But when finally, in quick, dizzying succession, we read repetitions of five key items . . .

> Stratum [I]: "leer la novela"; "su sillón favorito"; "terciopelo verde"; "su cabeza"; "los ventanales" ["read the novel"; "his favorite armchair"; "green velvet"; "his head"; "the great windows"] (9 [55])

. . . all in the last sentence . . .

> Stratum [II]: "la luz de los ventanales, el alto respaldo de un sillón de terciopelo verde, la cabeza del hombre en el sillón leyendo una novela" ["the

9. While discussing this story in the classroom, my student Sol Canal proposed a scenario according to which this sort of ending *would* be possible: The woman could be the secret author of the novel the man is reading, and the lovers' fastidious planning ensures that he will happen to be reading this novel when the man with the knife arrives to "fulfill the prophecy."

> light from the great windows, the high back of an armchair covered in green velvet, the head of the man in the chair reading a novel"] (11 [56])

. . . our minds can no longer keep at bay the persistent human disposition to make associations. Repetition has thus established the titular continuity.

Here I must reiterate that a connection between stratum [I] and stratum [II] is never spelled out by Cortázar. At most, continuity is a common assumption on the part of the reader. But why should that be the case? Since we are not provided any character names, why would we think that the man reading a novel in the beginning of the story is the same man who is reading a novel at the end of the story—especially as such a scenario would be physically impossible? Sure, concepts and items are repeated, but is not even a huge coincidence more likely than a character coming out of his tale to kill his reader, a thing that cannot actually take place?

It could be that one of the reasons we choose the impossible over the highly coincidental is that when reading fiction, we are not concerned about remaining within the bounds of what's possible. A more immediate reason we choose the impossible reading is that Cortázar's meticulous crafting *directs* us to it. Unlike in *At Swim* where we watch the metalepsis unfold (or, more accurately, twist) from above, in "Continuidad" we experience it as though we were in the characters' shoes. Like the worms on ground level, we are nudged along the tangle, and end up surprised and disoriented. At the end of the story, we ask, How did we get here? We thought we were on one narrative level and lo, we ended up on an incompatible other.

I have just traced the reader's steps as they pick up on textual clues that lead them to the idea that stratum [I] and stratum [II] have somehow intersected. Now let's consider where the reader's steps *begin*. Because the man on stratum [I] is described as doing the same thing as the reader of "Continuidad"—that is, he is reading a story—the reader's journey of "Continuidad" begins with a special kind of alignment with this man. The reader thinks, I am reading about someone reading; this man is reading a story, just like I am. In the process of following the trajectory of this man, the reader encounters what appears to be stratum [II], the trajectory of the couple in the man's novel. The reader, along with the man, now follows the couple's trajectory, which the reader assumes to be on stratum [II]. But in picking up on Cortázar's carefully placed repetitions the reader becomes aware of a shift: Seemingly out of nowhere, the two trajectories have intersected!

Phelan argues that "Cortázar's ending is effectively shocking because of how it transforms the relationships between the narrative and authorial audiences" ("The Mimetic" 167). He reasons, "For Cortázar's authorial audience, the illusion that the storyworld imitates the actual world gets shattered" (168). Now the reader can no longer identify with the stratum [I] man—and *will* not, since the reader does not want to share the man's end. Phelan writes that the illusion's shattering "suddenly opens up a huge gap between [the authorial audiences'] knowledge and beliefs and those of the narrative audience. By introducing this gap, Cortázar shifts the dominant mode of the story from mimetic to synthetic" (168). In moving from the mimetic orientation to the synthetic one,[10] the authorial audience goes from flying above the story—looking at the characters from a bird's perspective—to being as startled as the man in the green chair, experiencing the story "on the ground," like the worm.

Such a transformation in the relationship between audiences, and the consequent shift from mimetic mode to synthetic mode, does not occur in *At Swim. At Swim*'s reader need not go "down in the muck," is never nudged away from the dominance of the authorial-audience role. Whatever steps the reader of *At Swim* takes, they do not begin with the kind of identification engendered in "Continuidad" for a simple reason: In *At Swim,* the stratum [I] character is *writing* a story, rather than doing what the text's reader (and the stratum [I] character in "Continuidad") is doing, which is *reading* a story. The reader of *At Swim* may of course relate to the novel's stratum [I] character (the unnamed narrator) for a variety of reasons. But because this character is not exactly replicating the reader's activity of reading, we do not see that special kind of identification that involves the reader placing themselves in the character's shoes—on the ground, as it were. This is not to say that all examples of worm metalepsis feature a character who is reading; I am not ready to say that worm metalepsis relies on the reader's identification with the character, nor that reading about a character reading is the only way to forge identification. However, identification with the character certainly helps the reader maintain the mimetic illusion. In "Continuidad," the reader's special identification with the stratum [I] man—reading buddies!—greases the reader's route; they are more smoothly directed into a mimetic illusion, such that by the time they arrive at the counterintuitive realization, it is with great shock.

The shift of dominant from mimetic to synthetic happens because the reader takes on the character's perspective. A story such as "Continuidad," where this happens, is the kind that aligns with the counterintuitive characteristic of wormholes that is the subject of this chapter. Both the reader of this story and the individual inside the light cone find themselves accessible by, or with access to, an entity that, before the warp of metalepsis, or wormhole, was

10. See Phelan's definitions of these modes in "The Mimetic, the Synthetic, and the Criterion of Correspondence."

on the other side of an uncrossable boundary. Our F statement may therefore be rendered as

> [The storyfabric] curves such that {characters} on either side of an effectively uncrossable boundary can become connected.

Mapping Entailments of Warp

Bending Time

It is easy to disregard the "time" part of "spacetime" because the bending of time is difficult to imagine. Conjuring bent space is simpler in comparison; we can fold a piece of paper in our mind's eye with no trouble. Time dilation, on the other hand, is much less intuitive. The notion that proximity to a massive object can make time pass differently is bizarre; this is not something we witness on the scale of our lived experience.

Something similar is effected by ontological metalepsis—the "massive planet" in "Continuidad"—which makes the bending of time a metaphorical entailment of the Worm's-Eye-View Metalepsis || Spacetime Curvature fi-sci analogy. Among the various reader perceptions Cortázar shepherds, that of time may quietly be the most potent. Whether the reader picks up on Cortázar's maneuvers consciously or subconsciously, the author's steering through time serves to further solidify what the aforementioned repetitions accomplish. The following indicators of time are reproduced here in the order they appear in the story, spread out across the two perceived strata:

1. Stratum [I]: "Esa tarde" ["that afternoon"] (9 [55])
2. Stratum [I]: "el aire de atardecer" ["the air of afternoon"][11] (9 [55])
3. Stratum [II], it seems: "Empezaba a anochecer" ["It was beginning to get dark"] (10 [56])
4. Stratum [II], it seems: "la bruma malva del crepúsculo" ["the yellowish fog of dusk"][12] (10 [56])

With these clues of sequence—of time proceeding in more or less even intervals from afternoon to evening to darkening to dusk—we are nudged toward

11. Though in his translation Paul Blackburn chooses "afternoon," "atardecer" can translate to "evening" or "sunset"; thus, in the original Spanish, the move from phrase 1 to phrase 2 depicts a more gradated, sequential passage of time.

12. "Malva" means "mauve."

viewing the two strata as taking place consecutively, to perceive a continuity in time. But aren't these two strata like parallel lines, such that one should not be able to proceed from the other, whether in space or in time? Or, more in line with this chapter's analogy, like causally disconnected regions, such that one should not be able to connect to the other? How can one stratum continue into the other if they are ontologically different? It cannot: The two strata must therefore actually be one. Thus, the narrative's shepherding of the reader through a perceived continuity in time prompts and supports the reader's perception of a continuity in space.

This makes sense because, as the term *spacetime* suggests, though space and time are different kinds of dimensions, three of the former and one of the latter must all be taken into account to determine the state and path of an object. In conversation with Aldama, Phelan mentions "the way in which Cortázar works with time and space." In the space of stratum (I) "there is the time of [the man's] reading . . . and the time of the reading is the time of the read. . . . So what's happening between the lovers, and then the time of the male lover's traveling to that space [gets telescoped into the time of the read[13]]. . . . At the end, the time and the space converge in the interpenetration of the ontologies." Aldama responds, "The brilliance is in the way that Cortázar, we might say, dynamizes . . . or the dynamization of space and then the spacialization of time so that they're not of course separate, but they're controlled or manipulated, if you will . . . differently according to the devices, the shaping, the syntax that he uses" (45:03–46:55). Cortázar ushers readers down a warp in spacetime.

•

That we can felicitously substitute entities in the T expression with the corresponding entities in the F expression demonstrates that the same pattern is at play in both the fiction and the science, giving us the fi-sci analogy Worm's-Eye-View Metalepsis || Spacetime Curvature (see table 10). The counterintuitive in fiction—in this case characters from different ontological strata that somehow are able to interact—functions as an intermediary: Our ability to imagine this smooths the path to visualizing a phenomenon in physics that shares the same pattern. Our conceptualization of one character who is about to attack another character who exists on a different ontological stratum correlates to our description of how a wormhole makes it possible for causally disconnected regions to connect. The mental juxtaposition of these two

13. Phelan's addition, August 2024.

TABLE 10. Warp flowchart

	Phenomenon	Example	Expression
Science	1 Spacetime curvature	2 A wormhole can connect causally disconnected regions.	3 [Spacetime] curves such that {regions} on either side of an effectively uncrossable boundary can become connected.
Template	—	—	4 [A framework] curves such that {entities} on either side of an effectively uncrossable boundary can become connected.
Fiction	7 Worm's-eye-view metalepsis	6 "Continuidad de los parques," Julio Cortázar	5 [The storyfabric] curves such that {characters} on either side of an effectively uncrossable boundary can become connected.

phenomena (impossibly interacting characters and confoundingly connected regions) familiarizes the effects of spacetime curvature. The characters' bizarre contact was accepted under the set of assumptions we bring to fiction. Carrying this accepted shape over to relativistic physics helps to conventionalize the latter; the playful acceptance of a counterintuitive thing in fiction concretizes the similarly patterned counterintuitive thing in science.

CHAPTER 5

Ambiguity

A fundamental concept of quantum mechanics—and one of the theory's most baffling tenets—is superposition. In 1935 Schrödinger published what has become a famous imaginary scenario: a thought experiment whereby quantum superposition is applied to a classical system in the form of a cat in a box. The box consists of a radioactive atom, a detecting apparatus, a container of poison, and a cat. There is a 50 percent chance that the radioactive material will decay, which will lead to the detector's recording of a particle. According to the setup in the box, if the detector records a particle, the poison will be released and the cat will die (Gribbin 205; Charap 58–59). When we open the box, there is a 50 percent chance that we will find the cat dead, and a 50 percent chance that we will find the cat alive. The perplexity lies in this: Until we open the box and observe its contents, the cat exists in a superposition—he is both dead and alive.

Describing quantum superposition, Paul Dirac notes, "The original state must be regarded as the result of a kind of *superposition* of two or more new states, in a way that cannot be conceived on classical ideas" (12). Cannot be conceived indeed. How should we imagine a creature that is both dead and alive simultaneously, when the two states are constitutively opposite? In this chapter about nonclassical physics phenomena, we will find that accessing this seemingly impossible principle is made possible once we recognize its avatar in fiction.

Quantum Superposition

Why did Schrödinger devise this thought experiment, and in what context? John Gribbin holds that "Schrödinger thought up the example [of the cat] to establish that there is a flaw in the strict Copenhagen interpretation, since obviously the cat cannot be both alive and dead at the same time" (205).[1] John M. Charap describes the rationale for this thought experiment as follows: "Even though the rules of quantum mechanics were first formulated for systems on an atomic scale, there is no reason why they should not also be applied to macroscopic systems too. Indeed, they must be so applied if we are to accept that quantum mechanics is of general and universal validity" (58). The macroscopic system of Schrödinger's thought experiment is the box and its contents. Charap explains the controversy arising from Schrödinger's thought experiment:

> It is hard enough to imagine the state of the atom after an hour as being described by a wave-function which, until the act of measurement, is a superposition of two components, each with the same amplitude, one of which describes an undecayed atom while the other describes an atom which had decayed. But can we contemplate a wave-function describing such a superposition of dead cat and live cat? Are we really prepared to accept that until we open the box and "perform the measurement," there is some sort of superposition of dead cat and live cat inside the box, each occurring with equal amplitude after an hour? . . . How are we to think of ourselves as quantum systems? (59)

That the cat is both dead and alive simultaneously is counterintuitive, and difficult to wrap one's brain around. While Charap's questions have been discussed over decades, for our purposes it is enough to note that quantum superposition is a fact. Schrödinger's cat is an instance of fictionality, used to illustrate such a superposition on a classical scale. The cat is both dead and alive until someone opens the box and observes him. At the point of observation, the wave function—to use the language of quantum mechanics—collapses, and the cat takes on one state: Only then is the cat either dead or alive.

1. The "Copenhagen interpretation" refers to a "collection of ideas—uncertainty, complementarity, probability, and the disturbance of the system being observed by the observer" (Gribbin 121).

Quantum superposition can be conveyed in the S expression

A [quantum particle] is simultaneously {in state p} and not {in state p}.

Once we replace the science specifics "quantum particle" and "state," the T expression becomes

An [entity] is simultaneously {p} and not {p}.

Where in the realm of fictionality might we observe such a pattern?

Why the Cat Does Not Map

Schrödinger's thought experiment is an instance of fictionality, but it is one devised in order to illustrate that the pattern of quantum superposition cannot be mapped. Recall Gribbin's observation that "Schrödinger thought up the example to establish that there is a flaw in the strict Copenhagen interpretation, since obviously the cat cannot be both alive and dead at the same time"; Schrödinger wanted to show that the implications of the Copenhagen interpretation are nonintuitive at a classical scale. To use our terms, it is as though Schrödinger tried his hand at a fi-sci analogy so as to demonstrate that the pattern does *not* map. My goal differs from Schrödinger's; I want to show that the pattern *can* map. To do so, I will replace the existential duality at play in the thought experiment with a different kind of existential duality, one in line with this book's approach: a duality that hinges on fictionality.

The existential duality at play in Schrödinger's scenario is that of alive/dead; according to the thought experiment, the cat is in both of these opposite states at the same time. The reason Schrödinger succeeds in conveying the nonintuitive nature of superposition at a classical scale is that we indeed cannot imagine a cat that is *simultaneously* in both of those states—even if this cat is fictional. When we consider a fictional cat, we can certainly conceive of him alternating between the two states—alive, then dead, then fantastically alive again—something a nonfictional cat can't do. But what we cannot do is hold both of those perceptions at the same time. The simultaneity of incompatible states is at the heart of superposition's incomprehensibility. The cat does not offer us a path toward comprehending the simultaneity of incompatible states.

Now let's see if we can convey the simultaneity of incompatibilities using an existential duality that hinges on fictionality. The existential duality of fictionality is that of fictional/nonfictional. Instead of a cat facing the duality of alive/dead, we seek a cat—or other creature—facing the duality of fictional/nonfictional. And so we ask: What kinds of fictional creatures are simultaneously in these two conditions of fictional and nonfictional? Before we talk about a fictional being's condition, let's consider ontology—a branch of metaphysics concerned with the nature of being and existence—and see to what extent one can even speak of a fictional being's existence.

Ontology and Fictionality

The many connotations of "fiction" include dream, hallucination, falsity, impossibility, nonexistence, and social—rather than objective—construction. The connotation I wish to focus on is that of contrivance. This is in keeping with Phelan's aforementioned characterization of fictive discourse as "intentionally communicated invention, projection, or other means of directing an audience to consider nonactual states or events" (Gammelgaard et al. 7). In short, I will treat fictional entities as those that have been purposefully contrived—made up.

Fictional entities exist in that we interact with them, can speak of them, and are affected by them. Much has been written about the problem of the existence of fictional entities, as well as the nature of their existence. We can gain an interesting and useful point of departure from theories arrived at by the consideration of fictional characters. What are some of the characteristics of fictional characters, despite or even because of their fictionality? J. J. A. Mooij writes,

> It is nearly universally accepted that fictional entities (whatever they are) can be described, i.e., that correct statements can be made about them; other statements about them, of course, can be incorrect. The current standard view is that such statements are true or false according as they are conformable to, or incompatible with, something authoritatively said or implied in the relevant fictional text. (105)

For example, it is correct to say that Sherlock Holmes works closely with his colleague John Watson, because Holmes's contriver—Arthur Conan Doyle—authoritatively says so; it is incorrect to say that Watson is Holmes's nemesis since Watson consistently assists Holmes and since Doyle identifies another character, James Moriarty, as Holmes's antagonist. The correct description is compatible with the way in which Doyle imagined his fictional characters, and the incorrect description is incompatible with the fiction. Therefore, while

Holmes is indeed fictional, there are correct and incorrect things we can say about him, given what was "authoritatively said or implied in the relevant fictional text."

Additionally, Mooij delineates the nature of the existence of fictional entities: "Fictive objects do exist if, and only if, they have been *made* to exist, i.e. if, and only if, they have been created. . . . What would be necessary for real existence of a fictive entity would be some kind of substance or significance, or some kind of further articulation. That is, it should be anchored in the world that is already there" (113). Holmes has been created and made to exist, and he is articulated every time he appears in a book, show, and so on.

•

We have identified a precedent for sensically speaking about a fictional being's existence. Let's return to the original question: What kinds of fictional creatures are simultaneously fictional and nonfictional? What should our guiding F expression state? In keeping with Schrödinger's thought experiment let our alternative also be a living organism; when moving from the T expression to the F expression let's replace "entity" with "character." And since we are dealing with an existential duality of a fictional character, the adjective in brackets must be "fictional." The F expression is therefore,

A [character] is simultaneously {fictional} and not {fictional}.

Avatars of Quantum Superposition in Fiction

A Fiction That Is Not an Avatar

To be a close analog of quantum superposition, a scenario must meet the following criteria: [a] it must feature a duality of states; [b] the dual states must be incompatible; [c] the dual states must be simultaneously active. Schrödinger's cat meets [a] and [b] but not [c], since a cat cannot be simultaneously alive and dead.

Consider another candidate for avatar of quantum superposition: the language spoken by the alien Ariekei of China Miéville's *Embassytown* (2011). This language is referred to simply as "Language," and it features peculiar, fascinating characteristics. "To understand Language," writes Abigail Nussbaum, "Ariekei must hear it spoken by a sentient person who understands its meaning and imbues their speech with it. . . . because Ariekei speak simultaneously from two mouths, a single human can't communicate with them, nor can two people speaking in tandem, because the singular intent isn't there." Stan Carey elaborates, "Language, to be understood, needs a mind behind it." Language, sums up CD Covington, "requires two mouths, two voices. . . . And these two voices must come from the same person." To communicate coherently in Language, a speaker's one mouth speaks the Cut, and the other mouth speaks the Turn; crucially, the Cut and Turn must be spoken simultaneously.

Language meets [a] and [c] but not [b]. In requiring two voices that express different things, it features a duality of states, and because Language is spoken with two mouths that share a mind, these two states are issued and perceived simultaneously. However, the states in this duality are not incompatible—the thing said with mouth [1] can coexist with the thing said with mouth [2]; one voice does not preclude the other. In fact, the Cut and Turn *must* coexist for speech to have meaning; Ariekei do not comprehend one voice if it is unaccompanied by the other.

The Muppets

Where can we find fictional entities that simultaneously inhabit incompatible states, characters that are fictional and nonfictional at once? I submit that we find them sensically existing in ontologically incompatible realms: as fictional characters in the realm of fiction, like *Mr. Magorium's Wonder Emporium* and *The West Wing*, and as nonfictional characters in the realm of nonfiction, like *Vogue* and Instagram. Fictional entities that inhabit both scenarios are the Muppets.[2]

The nature of the Muppets' being—their ontological status—is in superposition because while they are fictional characters, they are often depicted as occupying the same ontological stratum as nonfictional beings. Rather than remain as fictional characters in a fictional setting, they continuously cross the ontological boundary and show up in nonfiction. Sometimes the Muppets appear in films, dramas, and sitcoms (where the setting is fictional); other times they appear in real-world events such as awards ceremonies or late-night talk shows (where the setting is nonfictional). Instances of the former include *The Great Muppet Caper, The Muppets Take Manhattan, Muppets from Space, Kermit's Swamp Years,* and Kermit the Frog's appearance on *30 Rock*

2. Kermit the Frog appears as a fictional character in *Mr. Magorium's Wonder Emporium*; Big Bird appears as a fictional character on season 5, episode 16 of *The West Wing* (Reid); Miss Piggy is interviewed by Lisa Niven-Phillips in British *Vogue*; Elmo has one million followers on Instagram ("Elmo").

("My Whole Life"). Instances of the latter include Miss Piggy joining Johnny Carson onstage to do a bit at the 52nd Academy Awards ("Johnny Carson"), Kermit's appearance on *The Ellen Show* ("Ellen Meets Kermit!"), and the Muppets joining Jimmy Fallon on his final show as the host of *Late Night with Jimmy Fallon* ("Andy Samberg").

In the context of *30 Rock*—a fiction—Kermit and Jack Donaghy are both fictional. But Kermit is more complicatedly fictional. He carries another layer of fictionality with him that Donaghy does not have. Whereas Donaghy is a fictional character played by nonfictional Alec Baldwin, Kermit is a fictional character playing himself, a fictional being. For the narrative audience of *30 Rock,* Donaghy is real, while Kermit is a fictional entity who can metaleptically interact with Donaghy. Announcing an Oscar nominee alongside nonfictional Carson, Miss Piggy is also involved in a metaleptical interaction, but this time the context is nonfiction. The playfulness of her "cameo" stems from this: She is presented as real like Carson, while paradoxically and inescapably retaining her fictionality.

Let's take a closer look at what happens when the Muppets show up in nonfiction. When they join Jimmy Fallon, they must be perceived, as far as the audience is concerned, as joining the same ontological stratum as Fallon—that is, the stratum of nonfiction. Otherwise, we are faced with the absurd situation of a nonfictional entity communicating with a fictional entity across the ontological divide. In order to make ontological sense of this situation, the viewers will likely ferry the Muppets onto the ontological stratum on which Fallon exists (all the while retaining awareness of the Muppets' fictionality) rather than the other way around, because the context—the entire show of *Late Night* and its participants—is one of nonfiction.[3] Even as the viewers make these cognitive adjustments they retain their understanding of the Muppets as ontologically different from Fallon: He's "just" real; they're simultaneously fictional and nonfictional. In other words, they are exemplifying superposition.

If this reasoning sounds familiar, that's because we just recently followed it. In the previous chapter I introduced an augmentation of the rhetorical model of audiences, proposing that a triple consciousness is in play when it comes to synontology. Applied to the *Late Night* example, this means that beyond viewing the Muppets as real while retaining the awareness that they are fictional, the audience also playfully disregards said awareness—all thanks to cognitive flexibility. Just as a *Star Trek* fan simultaneously holds two ontological perspectives when they make their pilgrimage to Captain Kirk's future

---

3. Though the argument that the two—Muppets and Fallon—must be perceived as joining the same ontological stratum would hold even if viewers were to ferry Fallon onto the ontological stratum of the Muppets and view *Late Night* as a world of fiction.

birthplace, so too does the audience of the Muppets on *Late Night*. The Muppets' synontology is thus an avatar of quantum superposition.[4]

The Muppets and Quantum Physics

A Muppets–quantum physics connection is not as farfetched as it may at first appear—and I do not say this because of the "Muppet Labs" skits featuring Dr. Bunsen Honeydew and Beaker. Rather, an affinity between the Muppets and the new physics is already recognizable in what Michelle Ann Abate argues is "a key facet" of *The Muppet Show*: nonsense (590). And lest it seem I am equating quantum physics with nonsense, allow me to explain.

The Muppets became famous in no small part because they "routinely embraced absurdity, irrationality, and foolishness. . . . Nonsense contributed to the show's mainstream success, its signature goofy humor, and—perhaps most importantly—its popularity with adults and children alike" (Abate 590). Some of the language Abate uses to portray the functions and power of nonsense in this series is similar to the language used to depict the paradigm shift brought on by the development of quantum theory in the twentieth century. Characterizing *The Muppet Show,* Abate describes a "world where . . . the laws of physics are commonly ignored" (598). Just qualify "physics" with "classical" and the same can be applied to how quantum mechanics comes across.

This is not at all to say that quantum mechanics is nonsensical. Throughout this book I have been characterizing elements of quantum mechanics as counterintuitive because its core ideas do not seem to align with our embodied reality. But quantum mechanics is not illogical; that we can describe it with mathematics cements it as logical. Quantum mechanics is not nonsensical; it bears internal consistency—makes sense in its context. In the realm of literary analysis, however, "illogical" and "nonsensical" are used in a different sense, and are perfectly applicable to the Muppets' behavior. The connection between quantum physics and the Muppets is not that they can both be described in the same terms but, rather, this: The counterintuitive in quantum

4. I would not go so far as to say that all synontological situations can function as avatars of quantum superposition. The brick-and-mortar Kwik-E-Marts and Reichenbach Falls plaque are isolated or stand-alone instances that are not frequent or widespread enough for people to routinely think of the store or Sherlock Holmes as ontologically ambiguous entities. The Muppets, on the other hand, and other entities such as Quidditch and the Wonka Bar, regularly pop up in both reality and fiction so that we might perceive them as fictional and nonfictional simultaneously.

physics has a similar effect to the nonsensical and illogical in the Muppets. Namely, in both cases, "the laws of [classical] physics are commonly ignored."

Vertical and Horizontal Axes

The Muppets examples mentioned earlier occur at what we may call a vertical axis: The state of the Muppets depends on the ontological condition of the setting in which they are featured. The ontological ambiguity of the Muppets operates on a horizontal axis as well: A Muppet's state can depend on the ontological states of *other Muppets*. The relationship between Big Bird and Mr. Snuffleupagus (Snuffy) illustrates ontological ambiguity on the horizontal axis.

Discussing the origin of Snuffy, Michael Davis writes that the show's producer Dulcy Singer and writer Tony Geiss "agreed that the character was always envisioned as Big Bird's imaginary friend, a fine idea while it lasted" (240). David Borgenicht explains why this did not last:

> It was 1985, and the writers had just about run out of ways for Snuffy to be "just missed" by the *Sesame Street* adults. Moreover, researchers had become worried about the message this was sending to children—that they might not be believed by the adults around them if they told them something unusual. . . .
>
> At that time, increasing numbers of stories about child molestation and abuse were being told in the media. The *Sesame Street* educators worried that if children saw that the grown-ups didn't believe what Big Bird said (even though it was true), they would be afraid to talk to adults about dramatic or disturbing things that happened to them—afraid that adults wouldn't believe them, either.
>
> So the writers and researchers decided that it was time to have Snuffy "become real" and be seen by the adult cast. (41)

Before Snuffy "became real," that is, until 1985, he was Big Bird's imaginary friend. For our purposes it makes sense to analyze that "version" of Snuffy's existence in our exploration of his ontological status.

Snuffy's status as an imaginary friend of Big Bird, who is himself a fictional character, complicates Snuffy's fictionality. As a fictional character fabulated by a fictional character, Snuffy is presented as a doubly fictional character—one who is twice removed from reality. But does that make him *more* fictional than Big Bird? Is he somehow less real, and can we even consider fictionality in terms of degree? Of course, it is true that the nonfictional Singer and Geiss fabulated both Big Bird and Snuffy. However, as part of their fabulation, they stipulated otherwise. Within the frame of reference of *Sesame Street,* the fabulation, creation, authorship of Snuffy is attributed to a fictional character—Big Bird. Once we enter the frame of reference of *Sesame Street,* the original fabulators, Singer and Geiss, disappear. This stipulation must hold for the narrative premise that Snuffy is Big Bird's imaginary friend to keep from crumpling. According to this stipulation the act of fabulating a character into existence has been repeated—but this time by a fictional being. The second-tier fictional—Snuffy—is presented as *more fictional* than the first-tier fictional—Big Bird—precisely because the act of Snuffy's fabulation is attributed to a fictional being, Big Bird. The repetition of the act has further diluted the "realness" of the new character—the second-tier fictional.[5]

The notion of attribution is the essential point when we consider how the second-tier fictional affects the status of the first-tier fictional. This is because the one attributed with fabulation is also by extension attributed with a theory of mind and creative capabilities.[6] This makes a fictional character who is attributed with fabulating another fictional character more like the author—and therefore more "real-like." Since Big Bird is presented as someone with a theory of mind and creative capabilities, it is easier to conceive of him—and therefore all other creatures on his ontological stratum (the other Muppets with the exception of Snuffy 1971–85)—as occupying the same ontological stratum as his fabulators (Singer and Geiss), who occupy the same ontological stratum as his audience.[7]

Other Dualities of the Muppets

In "'Starring Kermit the Frog as Bob Cratchit': Muppets as Actors," Ginger Stelle points to a different existential co-occurrence in the Muppets. She holds

5. If Snuffy were to have an imaginary friend of his own, this imaginary friend would be a further embedded, fabulated fictional character, and thus a third-tier fictional.

6. Of course, any character with consciousness would be presented as having a theory of mind and creative capabilities; they are like us—just fictional. What's notable about Big Bird is that his theory of mind and creative capabilities are foregrounded, highlighted each time the show's adults could not see Snuffy, which implied that Snuffy is imaginary—that Big Bird created him. Snuffy and his status as an imaginary friend of a fictional character were, to use Mooij's term, repeatedly articulated.

7. Per this chapter's argument, Big Bird and the other Muppets (with the exception of Snuffy before 1985) do not always occupy the same ontological stratum as their audience, but they can easily cross over from their fictional storyworld into the actual world.

that "the Muppets exist simultaneously as characters and performers" (93). Stelle writes,

> The Muppet movies represent the greatest display of this phenomenon as the films ask audiences to accept the Muppets on the same terms as human actors. . . . when Brad Pitt appears in a movie, audiences see him as the character he plays. The Muppet movies ostensibly ask audiences to see the Muppets as their on-screen characters rather than as their off-screen personae. However, the Muppets bring their characters with them. . . . the Muppets "play" themselves. (93)

This duality of character and performer is thus another duality that asks the audience to accept certain tenets, for instance: The Muppets exist as entities with lives and histories that are separate from the characters they play in the films in which they star. This places the Muppets on the same ontological stratum as the human actors alongside whom they perform. Stelle argues that *The Muppet Christmas Carol* "presents the most illuminating insights into the ability of the Muppets to become actors" (94). Kermit does not play "himself"; rather, he plays Bob Cratchit, just as a human actor would play Bob Cratchit in other adaptations of Charles Dickens's novella. Stelle notes that "at the same time, however, the Muppet-as-actor duality makes itself felt" and offers Gonzo—who plays Dickens—as an example: "Several of Gonzo's characteristic traits appear throughout the film. . . . He enjoys flying into the past attached to Scrooge with a grappling hook, something The Great Gonzo would enjoy. He also retains his love of chickens" (96).

In 2014 MSNBC anchor Rachel Maddow appeared as herself in the "Chapter 15" episode of the fictional Netflix series *House of Cards*, and in 2012 astrophysicist Neil deGrasse Tyson appeared as himself in a Superman comic ("Neil deGrasse Tyson"). The difference, however, between these examples and instances wherein the Muppets appear as themselves in any medium is that the Muppets originate in fiction, whereas Maddow and Tyson do not. The case of Muppets who take the role of actors who play characters in movies is different from the case of Muppets who appear in movies "as themselves." Examples of the latter include the aforementioned appearances of Big Bird on an episode of *The West Wing* and Kermit on *Mr. Magorium's Wonder Emporium*. Either iteration—Muppet as actor or Muppet as themself—presents a Muppet in superposition because the audience is invited to take into account the Muppet's concurrent "performance" at different ontological tiers.

Are the Muppets fictional, nonfictional, or both simultaneously? The short answer is that they are always fictional: They have been fabulated and never lose that dimension of their identity. But because there are contexts in which Muppets seem to be in ontologically paradoxical states, the fuller answer is a little more complex. We have two broad modes of context: (a) the Muppets in fiction; (b) the Muppets in nonfiction. In mode (a), the Muppets are your run-of-the-mill instances of fictionality. In mode (b), the Muppets' fictionality—while actually always there—is presented as fungible, or at least flexible. The audience of a talk show in which the host is interviewing Kermit about his breakup with Miss Piggy is playfully asked to be "in on the joke" that Kermit is a nonfictional guy like the host and the audience. The quirky humor of reading Miss Piggy's fashion advice in *Vogue* lies in the very fact that the audience is playing along with this bizarre ruse that Miss Piggy is nonfictional like they are. In mode (b), the audience is asked to watch/read as though the Muppets—who they know are fictional—are taking on the role of nonfictional beings. For this to work, two incompatible ontological statuses must be kept in mind simultaneously, creating a kind of ontological ambiguity. In mode (b), therefore, the Muppets illustrate superposition.

Mapping Entailments of Ambiguity

Quantum Entanglement

A metaphorical entailment of the Ontological Ambiguity || Quantum Superposition analogy—an additional property that maps over from source domain to target domain—is quantum entanglement. "If two particles interact at some point in time then the properties of these particles will remain connected at future times. A consequence of this is that determining the quantum state of one of the particles simultaneously determines the quantum state of the other particle, even if the two particles are a long way apart" ("Quantum Entanglement"). In our analysis the analogs of quantum particles are the Muppets; when we substitute the terms we obtain the following: If two Muppets interact at some point in time then the properties of these Muppets could remain related at future times. A consequence of this is that determining the ontological status of one of the Muppets simultaneously determines the ontological status of the other Muppet, even if the two Muppets are a long way apart.

Consider: At some point in time Kermit and Miss Piggy interacted in mode (a) on *Muppets from Space*. Across an ontological divide and many years later, Miss Piggy discusses her breakup with Kermit in mode (b) on a *Good Morning America* exclusive ("Miss Piggy Tells"). The property of these two Muppets that remains connected is that they exist on the same ontological stratum. Therefore, while Kermit does not take part in the *GMA* segment and is indeed apart from Miss Piggy, that she can present as a nonfictional being must mean that Kermit can present as such as well. Otherwise stated, the Muppets are entangled: The ontological status of one has implications for the ontological status of all.

In contrast to quantum entanglement, there is a property of quantum superposition that does not map onto ontological ambiguity, a property that cannot serve a metaphorical entailment: the collapse of the wave function. In the science domain, the resolution of ambiguity is possible. Once an entity external to the system interacts with the system to observe or measure it, the state of a quantum particle is set. No such resolution of ambiguity is possible for the Muppets in the domain of fiction. In mode (b), observation—the audience's interaction with the system—is actually the act that throws superposition into relief.

•

It turns out that the Muppets can offer a more robust illustration of quantum superposition than can Schrödinger's tabby. That's because the Muppets actually act out what it's like to simultaneously be in incompatible states. The cat cannot—that is why Schrödinger's is a successful thought experiment. That is, with the example of the cat Schrödinger succeeds in showing that the implications of the Copenhagen interpretation are counterintuitive at a classical scale. But though with the cat the pattern does not map, with the Muppets it does. We *can* actually hold in our minds the two notions that (1) the Muppets are fictional, and (2) (we are asked to pretend that) they are not.

We easily accept that fictional character Cookie Monster writes jokes for Jimmy Kimmel on *Jimmy Kimmel Live* ("Keira Knightley"). It is quirky and funny that Cookie Monster and Kimmel are somehow able to interact across the ontological divide. We are able to go along with the notion that the Muppets are ontologically ambiguous because of the rhetorical power of fictionality: When experiencing fiction we are primed to allow the seemingly impossible. Science has established that the confounding pattern "an entity is simultaneously p and not p" is mathematical reality. Fi-sci pattern mapping lets us take our acceptance of the pattern in fiction and apply it to our conceptualization of science.

It is both fascinating and fortunate that we can begin to glimpse quantum superposition by turning to another field entirely. The Muppets form a portal that allows conceptual access to a counterintuitive scientific phenomenon because their ontological ambiguity is analogous to a quantum particle's

TABLE 11. Ambiguity flowchart

	Phenomenon	Example	Expression
Science	1 Quantum superposition	2 An electron is simultaneously in two different states.	3 A [quantum particle] is simultaneously {in state p} and not {in state p}.
Template	—	—	4 An [entity] is simultaneously {p} and not {p}.
Fiction	7 Ontological ambiguity	6 Jim Henson's Muppets	5 A [character] is simultaneously {fictional} and not {fictional}.

occupation of two different states at once (see table 11). Fictionality—in this case, the unusual ontological niche of the Muppets—provides us with a schema to at least begin to imagine the confounding behavior of quantum particles.

And what of an introduction of cat and Muppet? It is only when we open the box and observe its contents and also establish the context of a given Muppet appearance that we will be able to determine which state of cat meets which state of Muppet.

CHAPTER 6

Recursion

According to the Decoherence Interpretation, when a quantum system interacts with an entity like a drop of water, the two become entangled, suppressing the property that sustains the superposition of states—the system's phase coherence. External interactions disrupt the quantum system such that it loses its "quantumness" (Schlosshauer 3). The idea that something else interfacing with a thing so affects this thing that it cannot maintain its unique properties is challenging to fathom. Yes, matter routinely transforms from one state to another, as when ice melts into water, or when water evaporates into water vapor. But phase change[1] involves changes in a substance's physical properties, like its density, its shape, and its intermolecular forces, not in its chemical composition, so its (chemical) identity does not change ("Teaching Matter and Phase Changes"). The suppression of phase coherence, on the other hand, dissolves the quantum system's ability to behave as a quantum system. This is a kind of change that is far more extreme and thus more challenging to conceptualize. Fear not, for there exists a wondrous narrative maneuver that can familiarize the strangeness: a true mise en abyme—an entity that contains a replica of itself within itself. I propose that this literary device has the capacity to disrupt a story's structural and ontological boundaries to such an extent that its "storyness" will come into question. Just as environmental entanglement fundamentally affects the quantum system, so mise en abyme fundamentally affects the story.

1. *Phase* in *phase change* is not related to *phase* in *phase coherence.*

Environmental Entanglement

Theoretical physicist Wojciech Zurek holds that "at the root of our unease with quantum theory is the clash between the principle of superposition . . . and everyday classical reality in which this principle appears to be violated" (4). Violated because the entities around us look like they're in particular states, rather than each in a superposition of states. There's a quantum realm where things work one way, and a classical realm where things seem to work another way, but how do the realms relate to each other? How do they connect? "The question of how the appearance of a classical world can be brought into agreement with, or even be derived from, the predictions of quantum mechanics extended into the macroscopic domain is one of the greatest foundational problems of quantum mechanics," writes physics researcher Maximilian Schlosshauer (8). He calls this "the problem of the quantum-to-classical transition" (6), that is, "the question of how effectively classical systems and properties around us emerge from the underlying quantum domain" (49).

Proposed by H. Dieter Zeh in a 1970 paper, a mechanism that came to be known as *decoherence* presents a solution to this problem. "Most physicists now accept," remarks Charap, "that it is the rapid decoherence of the quantum state for any complex system which is not well isolated from the environment that resolves the paradox of Schrödinger's cat, and, more generally, allows us to understand how classical behavior . . . can emerge from the underlying quantum mechanics" (99). So what is decoherence? To get there, let's first talk about coherence—phase coherence, to be precise. Charap explains, "When two waves remain 'in step' so that successive crests on each occur always in time with one another, they are said to be coherent. Coherence is a prerequisite for the characteristic wave phenomenon of interference, which occurs when two waves reinforce or cancel out where they are superposed or cross one another." He adds, "It is only when the quantum wave-function of two states are coherent that it is meaningful to talk of quantum superposition of those states" (200n11).

In *Quantum Interference and Coherence,* Zbigniew Ficek and Stuart Swain designate interference as the central enigma of quantum physics:

> Since the early days of quantum mechanics, interference has been described as the real quantum mystery. [Richard] Feynman, in his famous introduction to the lectures on the single particle superposition principle, referred in the following way to the phenomenon of interference: "it has in it the

> heart of quantum mechanics," and it is really "the only mystery" of quantum mechanics. (VII)[2]

With this endorsement we can reasonably regard phase coherence—the condition that enables quantum interference—as a fundamental characteristic of a quantum system, a property which gives the system its quantumness.

Now unless it is isolated, a quantum system interacts with the environment, and the ensuing entanglement disrupts the phase coherence of the system—in other words, it leads to decoherence. As McFadden explains, "Any open system is continually bombarded with photons, electrons, atoms and other particles. The quantum system will inevitably become entangled with the fate of billions of particles in its environment" (214). Interaction with the environment amounts to the environment *measuring* the quantum system (215). "Interference effects will be erased, so that the world will appear classical" (214–15). Zurek puts it this way: "Decoherence destroys superpositions. The environment induces, in effect, a superselection rule that prevents certain superpositions from being observed. Only states that survive this process can become classical" (21). Schlosshauer sums up how decoherence solves the problem of the quantum-to-classical transition:

> The coupling to the environment now *defines* the observable physical properties of the system. At the same time, quantum *coherence,* a measure for the "quantumness" of the system, is delocalized into the entangled system-environment state, which effectively removes it from our observation. This process . . . constitutes a key component in explaining how the classical world of our experience emerges from the underlying quantum substrate. (5)

•

Decoherence has two main consequences: "the environment-induced suppression of quantum interference and the selection of preferred 'pointer' states through the interaction with the environment" (Schlosshauer VIII). In this

2. The full quotation from Feynman's lecture is as follows:

> In this chapter we shall tackle immediately the basic element of the mysterious behavior in its most strange form. We choose to examine a phenomenon which is impossible, *absolutely* impossible, to explain in any classical way, and which has in it the heart of quantum mechanics. In reality, it contains the *only* mystery. We cannot make the mystery go away by "explaining" how it works. We will just *tell* you how it works. In telling you how it works we will have told you about the basic peculiarities of all quantum mechanics.

chapter, we'll explore the first consequence. I propose the following S expression to depict environmental entanglement's effect on quantum systems:

> [Environmental entanglement] disrupts the {phase coherence} of a <quantum system> such that the <quantum system> cannot maintain its (quantumness).

This expression includes details specific to the science. We replace these with nonscience terms so as to distill the expression into a generalized version that represents just the gist, rendering the pattern's shape. Replacing the words in the square brackets and the angle brackets is straightforward, but what are the roles of "phase coherence" and "quantumness" in the context of this expression? Since phase coherence is the thing that keeps a quantum system functioning as is, it serves as a preservative. Specifically, it allows the quantum system to maintain its unique properties. The T expression is therefore,

> [A mechanism] disrupts the {preservative} of an <entity> such that the <entity> cannot maintain its (unique properties).

Is there a narrative phenomenon that follows this shape, such that we could felicitously substitute the items in brackets with corresponding items in the realm of fiction? Where in fiction can we find a mechanism that disrupts the preservative of an entity to such a degree that said entity can no longer maintain its unique properties? Since our source domain is fiction, if we take the most macro viewpoint we can say this "entity" is a work of fiction—a story. Elements of a story that can influence the story—that is, affect how it unfolds—include the story's tropes or, more broadly, narrative maneuvers. As the terms themselves suggest, tropes (turns) and maneuvers (guides) indeed influence the course of a story, some so much so that they disrupt its structural and ontological boundaries, and because of them the reader is led to reconfigure their understanding of the story entirely. In the most extreme case, the narrative maneuver would cause such a disruption that a story's unique properties—the properties that enable it to "behave" as a story—may be dislodged. Let's therefore replace "preservative" and "unique properties" in the T expression with "structural and ontological boundaries" and "storyness," respectively, in the F expression. The F expression is shaping up to look like this:

> [A narrative maneuver] disrupts the {structural and ontological boundaries} of a <story> such that the <story> cannot maintain its (storyness).

To complete the F expression, we'll need to determine exactly which narrative maneuver fits this pattern.

Fiction || Physics

Before searching for the narrative maneuver that fits our F expression, let's zoom out and consider the fluctuation that opened a door to the more capacious version of our comparison, the fi-sci analogy family that we can call Fiction || Physics. This fluctuation is nothing less than a full paradigm shift that reverberated across multiple fields during the 1900s.

Paradigms

In his analysis of the history of science, Thomas Kuhn introduces the concept of a paradigm shift, whereby a change occurs in the fundamental assumptions underlying a science. Even though Kuhn limits the use of this term to the sciences, *paradigm shift* has since been applied to other disciplines as well. Indeed, a palpable permutation in the perception of the world, of one's reality, does not strike within the confines of one domain but rather is felt across domains. The same revolutionary and perhaps hitherto inconceivable mode of perception may appear in such distinct fields as biology, sociology, and art. An example of this is Charles Darwin's theory of evolution—with its basis in biology—and its subsequent manifestation in sociology in the form of social Darwinism. Another instance is psychoanalysis, expressed both in the ideas of Sigmund Freud and Jacques Lacan as well as in the art movements of Dada and Surrealism.

Most apropos to our present study is the manner in which a new mode of perception emerges in the ostensibly diametric fields of physics and literature. Patricia Waugh describes precisely such a paradigmatic shift in *Metafiction*:

> The historical period we are living through has been singularly uncertain, insecure, self-questioning and culturally pluralistic. Contemporary fiction clearly reflects this dissatisfaction with, and breakdown of, traditional values. Previously, as in the case of nineteenth-century realism, the forms of fiction derived from a firm belief in a commonly experienced, objectively existing world of history. Modernist fiction, written in the earlier part of this century, responded to the initial loss of belief in such a world. . . . The materialist,

> positivist and empiricist world-view on which realistic fiction is premised no longer exists. (6–7)

Waugh thus distinguishes between common perception and its corresponding literary form in the nineteenth century, and common perception and its corresponding literary form in the twentieth century. The older worldview to which she refers was shaken up in the realm of physics by the grand entrance of quantum mechanics early in the twentieth century. As we have seen in previous chapters, the implications of quantum mechanics have profoundly shifted our perception of reality. In what follows we will explore the ways in which Waugh as well as philosopher and semiotician Umberto Eco have discerned a patent analogy between these shifts in physics and the manifestation of the ensuing, altered worldview in fiction and reader response theory.

Metafiction and Quantum Physics

Metafiction is fiction that is self-referential: stories that tell of their own making, authors who show up in their own narratives, characters referring to their status as characters, and any element that draws the reader's attention to the constructed nature of the book they are reading. Waugh defines metafiction as follows: "fictional writing which self-consciously and systematically draws attention to its status as an artefact in order to pose questions about the relationship between fiction and reality" (2).

Metafiction reflects an awareness that "our knowledge of this world is . . . mediated through language" (3). The reason language serves a mediating function is that "language is an independent, self-contained system which generates its own 'meanings.' Its relationship to the phenomenal world is highly complex, problematic and regulated by convention" (3). We can understand mediation as a veil: This veil stands between us and objective reality, such that because of it we cannot clearly perceive objective reality. That is, mediation precludes objectivity. Waugh draws a parallel between "meta-ness" in fiction and "meta-ness" in quantum physics:

> In a sense, metafiction rests on a version of the Heisenbergian uncertainty principle: an awareness that "for the smallest building blocks of matter, every process of observation causes a major disturbance" (Heisenberg 1972, p. 126), and that it is impossible to describe an objective world because the observer always changes the observed. (3)

In attempting to describe (that is, write about) reality, we inescapably mediate our experience of it through language. By mediating (observing) reality, we construct (disturb) it. We can see that the principle that observation affects the object that is being observed is a pattern that quantum mechanics and metafiction share. Within our analytic schema of fi-sci pattern mapping, Waugh's mapping can be expressed as Metafiction || Uncertainty Principle.

The Open Work and Quantum Physics

Waugh directs this analogy between metafiction and the uncertainty principle toward an analysis of the *writer* and the *writing of* metafiction. In her book she refers to this writer as the "metafictionist"—one who shapes language in one's quest to represent the world. Shifting now to another stakeholder in the narrative system[3]—the *reader*—we unearth additional analogies between fiction and physics. The idea of the reader acting as an indispensable participator in the "result" of the text is explored by Eco in *The Open Work*. Eco focuses specifically on modern art forms, and defines an "open work" as one that requires, as part of its eventual formation, the audience's participation. He grants that "any work of art, even if it is not passed on to the addressee in an unfinished state, demands a free, inventive response, if only because it cannot really be appreciated unless the performer somehow reinvents it in psychological collaboration with the author himself" (4). However, Eco distinguishes between the requisite openness inherent in any art, and the particular openness that is a characteristic of only certain kinds of works, ones that are "open" in a far more tangible sense: "We can say that they are quite literally 'unfinished': the author seems to hand them on to the performer more or less like the components of a construction kit" (4).[4]

These modern, novel kinds of works coincide with a modern trend in science; Eco, like Waugh, points out that art and science move together within larger paradigm shifts. Eco cites the examples of Stéphane Mallarmé's *Livre* [The Book] and James Joyce's works, particularly how the former relates to non-Euclidean geometry and how the latter relates to the spatiotemporal continuum (Eco 14). He writes, "In every century, the way that artistic forms are structured reflects the way in which science or contemporary culture views reality" (13). Eco provides yet another instance where the terminology of physics is used to discuss art: "[Henri] Pousseur has offered a tentative

3. The narrative system is the arena wherein narrative is produced and consumed.

4. As examples, Eco cites musical works by Karlheinz Stockhausen, Luciano Berio, Henri Pousseur, and Pierre Boulez.

definition of his musical work which involves the term 'field of possibilities'" (14).

From physics in general, Eco makes his way to quantum mechanics and Heisenberg in particular:

> Perhaps it is no accident that these poetic systems emerge at the same period as the physicists' principle of *complementarity,* which rules that it is not possible to indicate the different behavior patterns of an elementary particle simultaneously. To describe these different behavior patterns, different *models,* which Heisenberg has defined as adequate when properly utilized, are put to use, but, since they contradict one another, they are therefore also complementary. Perhaps we are in a position to state that for these works of art an incomplete knowledge of the system is in fact an essential feature in its formulation. (15)

Eco is comparing modern literature—the "open work"—and quantum mechanics in the sense that in both, "incomplete knowledge of the system is . . . an essential feature in its formulation," as well as in the sense that the text has, like quantum mechanics, elements of "indeterminacy, complementarity, noncausality" (252n9). Within our schema of fi-sci pattern mapping, Eco's comparison can be expressed as Open Work || Indeterminacy/ Complementarity/Noncausality.

There might be one more fi-sci pattern in Eco's study. Arguably, even though he does not specifically name it as such, Eco evokes the wave function collapse in his discussion of performance:

> Every performance *explains* the composition but does not *exhaust* it. Every performance makes the work an actuality, but is itself only complementary to all possible other performances of the work. In short, we can say that every performance offers us a complete and satisfying version of the work, but at the same time makes it incomplete for us, because it cannot simultaneously give all the other artistic solutions which the work may admit. (15)

This can be understood to mean that every performance constitutes a collapse of the wave function. All other readings of the piece ("artistic solutions which the work may admit") are potentially possible before it is performed, but the act of performance actualizes the work and all other possible readings thereof "disperse" (at least, in that moment of measurement/ observation). Within our analytic schema, we might express this pattern as Performance || Wave Function Collapse.

An Avatar of Environmental Entanglement in Fiction

A New Pattern

Waugh's and Eco's analyses give us two—possibly three—fi-sci analogies that we can collect under the umbrella of the Fiction || Physics analogy family. In this chapter we are exploring a potential kindred analogy that would be at home under this umbrella, and thus far we have arrived at a formulation of the F expression that is missing just one piece—a specific narrative maneuver that would replace the general "A narrative maneuver" currently in the square brackets:

[A narrative maneuver] disrupts the {structural and ontological boundaries} of a <story> such that the <story> cannot maintain its (storyness).

Reconfiguration

The pantheon of literary devices, tropes, and maneuvers is vast; its members impact stories in various ways and to different degrees. The physics can help us pinpoint the qualities of the particular maneuver we seek, one that would fit into the square brackets and complete the F expression. Recall Charap's account of phase coherence: "Coherence is a prerequisite for the characteristic wave phenomenon of interference, which occurs when two waves reinforce or cancel out where they are superposed or cross one another. . . . It is only when the quantum wave-function of two states are coherent that it is meaningful to talk of quantum superposition of those states" (200n11). This quality of phase coherence that allows it to keeps a quantum system behaving as a quantum system is what led us to characterize phase coherence as a preservative. In the F expression, "preservative" is replaced with "structural and ontological boundaries"; these are indeed the qualities of a story that give the story its particular "coordinates." In the science, the mechanism of environmental entanglement has the capacity to disrupt the preserving phase coherence. The maneuver we seek should likewise have the capacity to not only nudge a story from its perch but alter its coordinates—enough to prompt a reconfiguration of the story.

By definition, narrative devices, tropes, and maneuvers guide readerly response to, engagement with, and perception of a story, but not all of them do so in a way I'd call disruptive. Red herrings, MacGuffins, and deus ex machina guide readerly response and engagement, but they typically fall short of prompting significant reconfiguration. My view and use of *reconfiguration* follows Yonina Hoffman's conceptualization of the term. Building on the framework of configuration and coherence Rabinowitz presents in *Before Reading,* Hoffman sees "*reconfiguration* operating in between" the two, "forward-looking because it is the formation of a working model of a narrative experienced in-process, but backward-looking because it is the retrospective reorganization of textual details in light of disruptive new information" (440). Among the most disruptive of disruptive new information is delayed disclosure about narrative frames.

Gerald Prince defines "frame narrative" as "a narrative in which another narrative is embedded; a narrative functioning as a frame for another narrative by providing a setting for it" (33). What kind of impact does a frame have? If you picture frame impact as a spectrum, on the one end you would find frames that are informative. To be sure, even "merely" informative frames can change a story's coordinates. Abbott demonstrates this in his discussion of the frame narrative in Henry James's *The Turn of the Screw.* "Framing narratives," Abbott writes, "can, and often do, play a vital role in the narratives they frame" (*Introduction to Narrative* 30). Sometimes this role is small, and other times it is considerable. In the case of James's novella, the framing narrative has "a significant effect on the embedded narrative. Shorn of its framing narrative, *the narrative of the governess is not the same narrative it is when embedded*" (30, emphasis added). The narrative of the governess is one thing when framed, and a different thing when not framed.

It seems to me that informative frames whose main effect is to contextualize the narrative would be part of a story's setup, and as such would typically be revealed early on. One might say that generally, early disclosures about frames would aid the reader's configuration of the narrative. Toward the other end of the spectrum are frames that we might characterize as subversive. Disclosures about these frames would be made much later on in the narrative, at a point where they would trigger *re*configuration. Consider Ian McEwan's *Atonement,* which prompts a substitution of frames very late in the novel—one that significantly changes how a reader sees the story: Upon experiencing this substitution, the reader must re-evaluate how the story's narrative levels relate to each other, and must also therefore reassess narrators, characters, their motivations, their actions—even the notion of truth itself. Delayed disclosures about frames such as this are jarring because by and large the later they are issued, the more reconfiguration they demand. This can be disruptive to such an extent that, as we see in *Atonement,* delayed disclosure about frames can have a substantial impact on the story, turning upside down what we think of as fiction within the fiction, at times even leading readers to see the story anew.

"Subversive frame shifts" and "belatedly disclosed frames," then, both seem to be good candidates for the narrative maneuver that would fit in the F expression's square brackets. Still, I propose we continue even further along the frame-impact spectrum. What we gleaned from reviewing the effect of environmental entanglement on a quantum system is that we should seek not a clip but a full-fledged collision: We need a kind of frame shift that is even more powerful than McEwan's delayed disclosure. Focusing on the first adjective in Hoffman's "disruptive new information," my sense is that mise en abyme is the best choice: It may just be the most disruptive genus in the frame family of narrative maneuvers. It prompts a kind of reconfiguration not offered by any other framing device—or, to my mind, by any other maneuver.

THE TRUE MISE EN ABYME

Mieke Bal refers to the outer and inner texts in a framed narrative as primary and embedded fabula, respectively. There can be several relationships between primary and embedded fabulae; the relationship relevant to our purposes is one whereby "the embedded story . . . resembles the primary story" (58). Bal specifies, "This phenomenon (the embedded text presenting a story that resembles the primary fabula) is comparable to infinite regress. In French the term is 'mise en abyme'" (62). As I shall soon argue, the qualification "resembles" does not suffice for what I consider a *true* mise en abyme.

Framed and embedded narratives do not always create mise en abyme. Even when the embedded story resembles the frame story, we do not necessarily get recursive mirroring. Mise en abyme is at hand in a text that is not merely mirrored but rather mirrored back and forth ad infinitum. The anchor of the mise en abyme is the point where the primary fabula and embedded fabula converge; a story within a story without a convergence remains a framed portrait, and, at most, a one-way reflection. Moreover, a mere convergence between two diegetic levels does not guarantee a mise en abyme. The mise en abyme is formed only when both converging elements are mirrored, that is, mirrored recursively. Here, a dizzying example from—who else?—Borges: "Esa noche que está en el centro de las 1001 Noches, cuando la reina Shahrazad . . . se pone a referir textualmente la historia de las 100 Noches, con riesgo de llegar otra vez a la noche en que la refiere, y así hasta lo infinito" ["That night at the centre of the *1001 Nights,* when the queen Scheherazade . . . begins to retell, verbatim, the story of the 1001 Nights, with the risk of returning once again to the night on which she is telling it—and so on, *ad infinitum*"] (*Obras completas, 1923–1972* 477 [Hurley 125]). The convergence opens the vault to the abyss, an infinite recursion whereby information is endlessly repeated.

I base my criteria for a true mise en abyme on a study of the term's beginnings. The term *mise en abyme* originally comes from a diary entry by André Gide, reproduced in Lucien Dällenbach's work, and in part below:

> In a work of art, I rather like to find thus transposed, at the level of the characters, the subject of the work itself. . . . in literature, there is the scene in which a play is acted in *Hamlet*; this also happens in many other plays. In *Wilhelm Meister*, there are the puppet shows and the festivities in the castle. In *The Fall of the House of Usher*, there is the piece that is read to Roderick, etc. None of these examples is absolutely accurate. What would be more accurate . . . would be a comparison with the device from heraldry that involves putting a second representation of the original shield "en abyme" within it. (Dällenbach 7)

Dällenbach makes the following interpretations: "The *mise en abyme*, as a means by which the work turns back on itself, appears to be a kind of *reflexion*"; "what fascinated [Gide] must have been *the image of a shield containing, in its centre, a miniature replica of itself*"; "*a 'mise en abyme' is any aspect enclosed within a work that shows a similarity with the work that contains it*" (8). Dällenbach continues to define the term, "the relation of the narrator N to his/her story S being the same as that of the narrator/character n to his/her story s" (18). This definition of mise en abyme indeed suggests a mirroring and reflexivity. Furthermore, the concept of mise en abyme must be considered in terms of infinite recursion. Since the miniaturization is, as Dällenbach writes, "a miniature replica" of the shield, the miniaturization should also in turn include a miniature replica of itself. If this were not the case, the reproduced miniature would not be a precise reproduction of the surface of the shield. The miniature would be a small version of *another, different* shield. In order for it to be a precise reproduction, it would have to include a reproduction of itself ad infinitum.

This recursive—as opposed to one-time reflexive—interpretation of mise en abyme would preclude works that might be cited as involving mise en abyme because of their containing a related embedded fabula within the primary fabula. Examples of such stories are Edgar Allan Poe's "The Fall of the House of Usher" and Margaret Atwood's *The Blind Assassin*. In these instances the embedded fabula is related to the primary fabula, and it may mirror it in terms of the subject matter or the characters' attributes, but the reflection only happens once instead of recursively. Otherwise stated, there is no second mirror in the narrative that is held up to the first mirror.

We can see that there are varying interpretations of what entails a narrative mise en abyme. For Bal it suffices that the inner fabula resembles the outer fabula. Dällenbach is similarly content when "any aspect enclosed within a work . . . shows a similarity with the work that contains it." Brian McHale sees it as follows:

> A true *mise-en-abyme* is determined by three criteria: first, it is a nested or embedded representation, occupying a narrative level inferior to that of the primary, diegetic narrative world; secondly, this nested representation *resembles* (copies, says Hofstadter) something at the level of the primary, diegetic world; and thirdly, this "something" that it resembles must constitute some salient and continuous aspect of the primary world, salient and continuous enough that we are willing to say the nested representation *reproduces* or *duplicates* the primary representation as a whole. (124)

To call a representation mise en abyme, Bal, Dällenbach, and McHale do not require a literally identical copy that recurs infinitely. For them, similarity or a partial recursion suffices, so long as, at least for McHale, the "something" that the nested representation resembles is "salient and continuous enough." In my view, however, there should be a higher bar to clear. A true mise en abyme is a representation that contains a complete replica within itself, forever.

This higher bar can be cleared by certain visual depictions of recursion. Infinite recursion is sometimes called the Droste effect, named after the Dutch company that since 1904 has been selling cocoa in a tin featuring an illustration of itself within itself. Because it is not possible—by definition—to complete an illustration depicting infinity, we settle for an approximation of infinite recursion. But while the illustration functionally depicts at most two recursions of itself, the *implication* is that the image recurs forever. A photograph of the reflection inside a mirror that is facing another mirror is probably the closest we can come to depicting infinite recursion in a static image, but even here we arrive at a limit, past which the eye can no longer detect additional recursions. Like the Droste tin, the most such a photo can do is *suggest* infinite recursion. This matter of suggestion applies in narrative mise en abyme as well. Imagine mise en abyme as a three-dimensional, never-ending spiral. The kind of nesting assembled by mise en abyme undeniably implies that each successive loop is further embedded in its totality.

The recursive spiral is inherent even if we do not actually follow down its path.

These implications and suggestions of infinity can be disruptive to different degrees. Where mise en abyme forms a relatively small, local recursion—one that features in the course of the story but does not encompass the entire story—the impact on how we read the main story, and on the story's coordinates, is minor. Imagine, say, a whodunit novel, wherein over the course of the investigation a detective comes across a text written by one of the suspects that goes like this: "It was a dark and stormy night. The crew said to the captain, 'Captain, tell us a story.' The captain said, 'It was a dark and stormy night'" (Coolidge et al. 547). This fragment certainly sets up an infinite loop: "It acts as an instruction that calls another instance of itself," with the implication that this calling will go on ad infinitum (547). However, the recursion functions as an "isolated" occasion within the greater narrative; it does not serve to embed the greater narrative within itself, and as such it does not occasion the reader to reconfigure the greater narrative. The main story frames the fragment, but this framing is at most informative, contextualizing. This kind of local recursion—even if it pops up late in the novel—cannot be said to disrupt the novel. I am more interested in the global mise en abyme, by which the main story—the whole story—is revealed to act as "an instruction that calls another instance of itself." As I will return to shortly, it is mise en abyme at this scale that prompts a staggering disruption of the story.

With the distinction between local and global mise en abyme in mind, and with our focus being on the latter, here is what I believe sets this narrative maneuver apart from all others: Mise en abyme is distinct because it alone has the capacity to continually add new material to a story, endlessly expanding it. This adding of new material is different from a reader's ability to expand on a storyworld by engaging with its characters and settings, which is part and parcel of experiencing a work of fiction.[5] All narratives allow one to keep adding details in one's mind, thereby extending the storyworld. But mise en abyme alone is a built-in, self-contained expander. The mechanism of infinite recursion is remarkable because here a finite set of words implies infinite new information: The point of convergence triggers a never-ending cycle of added material. That is more material than can ever be contained between any

5. As part of this extended engagement, readers reason about characters' motivations; wonder about characters' lives beyond what's described on the page; imagine possible pasts, possible futures, and unexpressed thoughts. (Though a closer look at how readers relate to and think about characters is beyond the scope of the present chapter, both rhetorical and cognitive approaches to narrative explore such readerly engagement with characters.) The reader may also elaborate on storyworld aspects other than characters, such as culture, cosmology, or the built environment.

book's covers—more material than can ever be written. One might argue that the information derived from an infinite recursion is not new information, that text that repeats over and over again only provides a greater amount of the same information. But this is incorrect: As I will explain with more detail later in this chapter, the story on each of the endless spiral's layers differs from its predecessor in that it is embedded in one additional frame, which makes the material on every layer unique. Each of the embedded recursions in the never-ending spiral adds new information—new material—to the story. Thus perpetually expanding, the story leaves its original coordinates in the interstellar dust.

Especially Disruptive New Information

The unique mechanism of mise en abyme is intriguing, but what makes it disruptive? In what way does the discovery of infinite recursion invite a kind of reconfiguration not prompted by any other narrative maneuver? As I indicated above, among narrative maneuvers mise en abyme alone offers endless new information. And while this may seem to quite literally be a plus, it can arguably affect readerly response in the form of a minus. What's withdrawn at the moment the reader reaches the point of convergence is the possibility of an ending—that component Frank Kermode argues is crucial to our understanding of stories—whether those in fiction or our own: "In 'making sense' of the world we still feel a need . . . to experience that concordance of beginning, middle, and end" (35).

Kermode turns to the ticking of a clock: "*Tick* is our word for a physical beginning, *tock* our word for an end. We say they differ. What enables them to be different is a special kind of middle. We can perceive a duration only when it is organized" (45). Continuing with this framework he holds that we "need . . . fictional devices" "to maintain the experience of organization" in a novel. "To maintain within that interval following *tick* a lively expectation of *tock*, and a sense that however remote *tock* may be, all that happens happens as if *tock* were certainly following. All such plotting presupposes and requires that an end will bestow upon the whole duration and meaning" (46). If all that happens happens as if *tock* were certainly following, what happens when we suddenly become absolutely certain that no *tock* is forthcoming, that there will be no end to bestow upon the whole duration and meaning?

Having inferred that the embedding goes on forever, the reader understands that in some sense the story can never be wrapped up. Of course, the work of fiction ends in that the reader can reach the end of the book. But when—even conceptually—going down the route presented by the newly discovered spiral, when considering the tale that is abysally embedded, the reader is denied any form of conclusion: The story will forever circle its way back to yet another point of convergence. This feeling of being left hanging may be somewhat similar to what readers experience when, at the end of *The Turn of the Screw*, they realize there will be no return to the frame narrative. Still, the lack of symmetry and the uneasiness this may engender do not come close to the realization that resolution and closure are mathematically beyond reach. To illustrate the severity, let's turn to Borges once more. When last we met his words, I was quoting from "El jardín de senderos que se bifurcan" ["The Garden of Forking Paths"]. The narrator characterizes Scheherazade's retelling of the story of *1001 Nights* as a risk: "the risk of returning once more to the night on which she is telling it—and so on, *ad infinitum*." Why is this risky?

Borges returns to this very mise en abyme in his essay "Magias parciales del Quijote" ["Partial Magic in the Quixote"], with the following description: "La noche DCII, mágica entre las noches. En esa noche, el rey oye de boca de la reina su propia historia. Oye el principio de la historia, que abarca a todas las demás, y también—de monstruoso modo—, a sí misma" ["The six hundred and second night, magical among all the nights. On that night, the king hears from the queen his own story. He hears the beginning of the story, which comprises all the others and also—monstrously—itself"] (*Obras completas, 1923–1972* 668–69 [Irby 195]). Why is this monstrous?

On the last page of "Magias parciales del Quijote," Borges offers an answer.

> ¿Por qué nos inquieta que el mapa esté incluido en el mapa y las mil y una noches en el libro de *Las Mil y Una Noches*? ¿Por qué nos inquieta que Don Quijote sea lector del *Quijote*, y Hamlet, espectador de *Hamlet*? Creo haber dado con la causa: tale inversiones sugieren que si los caracteres de una ficción pueden ser lectores o espectadores, nosotros, sus lectores o espectadores, podemos ser ficticios.
>
> [Why does it disturb us that the map be included in the map and the thousand and one nights in the book of the *Thousand and One Nights*? Why does it disturb us that Don Quixote be a reader of the *Quixote* and Hamlet a spectator of *Hamlet*? I believe I have found the reason: these inversions suggest that if the characters of a fictional work can be readers or spectators, we, its readers or spectators, can be fictitious.] (*Obras completas, 1923–1972* 669 [Irby 196])

But there's yet another reason why these infinite recursions may disturb us. In addition to having misgivings about our own ontological status, we may feel bereft of a conclusion. Ann Jefferson cites Borges's six hundred and second night to illustrate what she sees as the "one irremediably disruptive form that *mise en abyme* can take," disruptive because it "short-circuits the entire narrative" (197). "As far as narrative theory in general is concerned," she writes, the example of that night is "disturbing because it precludes an end. What looks initially like a progression towards an anticipated conclusion is blocked by an endless return to the beginning." Due to the infinite recursion, "the narrative is prevented from carrying out its ultimate purpose which is to end" (198). Mise en abyme is risky and monstrous because while the spiral continues endlessly down the void, the story enclosed therein is void of an ending.

When faced with the monstrosity of infinity, readers may reach for closure by thematizing the recursion. In the case of the six hundred and second night, this can take the form of moving one's focus from the story's (now unattainable) end to what the recursion itself might bring to light. For example, as Borges suggests, it may lead us to wonder whether we ourselves are fictional characters. Or it might foreground the power of narrative to effect change. The reader needs an end to such an extent that if they can't get one to the story, they will try to approximate the resolution an end provides in another way. This illustrates another disruptive facet of mise en abyme: It will incite the reader to be creative in how they deal with narrative "deficits."

It isn't unusual for readers to encounter all manner of elisions that are never bridged. But a story that lacks a conclusion not because it ended ambiguously, or prematurely, but because, by design, by virtue of a built-in, self-contained mechanism it will never end, is anomalous in that it undermines the shape we typically ascribe to stories. A local mise en abyme, while still a built-in, self-contained mechanism of infinite recursion, does not undermine the tripartite cohesion of beginning, middle, end (in whichever order they may unfold); when the recursion is an isolated fragment within the larger story, it does not pre-empt—because it does not access—the larger story's end. But in the case of global mise en abyme the end is revealed to be forever unreachable. This evaporation of an ending is the most disruptive kind of new information; it invites a particularly substantial reconfiguration because when faced with mise en abyme, the reader must reassess their very notion of "story."

While maneuvers like metalepsis and belatedly disclosed frames can disrupt a story's structural and ontological boundaries and bump it off its coordinates, mise an abyme alone has the capacity to dislodge a (the?) property that enables a story to "behave" as a story. In that this mechanism precludes the possibility of an ending, the sought after and essential *tock*, it may very well rid the story of its storyness. Other tropes and devices might work for the first half of the F expression, but they are not so disruptive as to affect a story's unique properties. Mise en abyme is the narrative maneuver that is analogous to environmental entanglement, the puzzle piece that fits into the F expression's square brackets:

> [Mise en abyme] disrupts the {structural and ontological boundaries} of a <story> such that the <story> cannot maintain its (storyness).

Cien años de soledad

A story in which we can find an avatar of the environmental entanglement pattern—one that features a true, global, all-encompassing mise en abyme—is Gabriel García Márquez's *Cien años de soledad* [*One Hundred Years of Solitude*]. This novel provides an ideal example of infinitely regressive mirroring: Its last three pages not only set up a mirror but place it in front of a second mirror. Deciphering Melquíades's parchments, Aureliano Babilonia realizes that they reveal "la historia de la familia, escrita por Melquíades hasta en sus detalles más triviales, con cien años de anticipación" ["the history of the family, written by Melquíades, down to the most trivial details, one hundred years ahead of time"] (García Márquez 469 [Rabassa 421]). Aureliano Babilonia reads about his birth and life, eventually reading about the very moment in which he is reading about his reading it: "Aureliano . . . empezó a descifrar el instante que estaba viviendo, descifrándolo a medida que lo vivía, profetizándose a sí mismo en el acto de descifrar la última página de los pergaminos, como si estuviera viendo en un espejo hablando" ["Aureliano . . . began to decipher the instant that he was living, deciphering it as he lived it, prophesying himself in the act of deciphering the last page of the parchments, as if he were looking into a speaking mirror"] (470–71 [422]). The mirror into which Aureliano Babilonia is looking is the novel we have been reading. The text in the book we are holding is identical to the text in the parchments written and narrated by Melquíades. The book we are holding is embedded (as the parchments) in the already embedded (for us) world of Macondo. In our segment of the mise en abyme spiral we are reading the same story that Aureliano Babilonia is reading in *his* segment of the spiral.

The infinite recursion in *Cien años* is as follows: Aureliano Babilonia is reading a narrative in which he reads about himself reading this narrative, in which he reads about himself reading this narrative, and so on. Aureliano Babilonia, in reading that he is reading the parchment that is the embedded fabula, is holding up a mirror to the mirror that is the primary fabula—which contains his reading of the embedded fabula. As Julio Ortega points out, Aureliano Babilonia "is the reader who ends up being the sign of his own reading" (11). This situation is an especially complex version of shifting frames because here it turns out that the embedded fabula *is* the primary fabula.

André Brink cites Michael Wood's discussion of loops in *Cien años*—situations "through which it becomes possible to look both forward and back in the same moment" (Brink 245). One instance of this is that "by the time Meme's son is brought home (p. 243), we have already learnt about his arrival (p. 239)" (246). Brink continues, "Ultimately, all these loops serve as a *mise-en-abyme* of the overall narrative situation: the history of the Buendías as the recapitulation of a story already written a hundred years before by Melquíades" (248). Because the primary fabula is recapitulated in whole by the embedded fabula, infinite recursion can occur: The novel mentions a manuscript, which, in telling the story of the novel, will mention a manuscript, and so on, forever.

Cien años contains other instances of recursion. There are the examples of perennial creation and destruction: The goldfish are formed from the gold only to be melted and formed again; the shroud is woven only to be unwoven and woven again. These recursions reflect the motif of solitude. However, the only time the embedded fabula (the parchments) holds up a mirror to the mirror of the primary fabula (the novel) is when Aureliano Babilonia reads about himself reading. Infinite recursion occurs because two mirrors are reflected in each other. The embedded fabula is revealed to be a mirror that mirrors the primary fabula, which in turn reflects back to the embedded fabula, recursively; the embedded fabula reflects a part in the primary fabula that leads *directly* back to the embedded fabula.

Into the Abyss

In mise en abyme, the framing winds recursive loops into an infinite spiral. I had suggested we imagine the spiral as three-dimensional so as to account for an abyss's depth. Each time a reading is applied to the text at the point of recursion, the reader moves into the spiral's next loop; each additional loop adds new information. Earlier, I noted that the text on each of the loops, or layers, differs from its predecessor in that it is embedded in one additional frame. That's because in each successive loop the frame of reference has shifted. The story is the same in terms of the signifier (the letters on the page do not physically change) but different in terms of the signified because it is now *embedded within the story of the previous reading.*

Let's closely consider an example from *Cien años*. We'll call the starting point of the primary fabula—the beginning of the novel that the reader is holding—point [1]. The text at that point states, "Muchos años después, frente al pelotón de fusilamiento, el coronel Aureliano Buendía había de recordar aquella tarde remota en que su padre lo llevó a conocer el hielo" ["Many years later, as he faced the firing squad, Colonel Aureliano Buendía was to remember that distant afternoon when his father took him to discover ice"] (9 [1]). Point [2] is located at the end of the novel, which is the moment when the first-tier loop is completed. It represents the instant in the text at which Aureliano Babilonia finally decodes Melquíades's prophetic parchments. Point [2] is where Aureliano Babilonia, having deciphered the parchments, reads "Muchos años después . . ." ["Many years later . . ."]. This is the point of convergence between the two tiers—the two diegetic strata.

Point [2] is therefore the place where the second loop of the spiral begins. If the reader were to follow Aureliano Babilonia's reading, they would *re-enter* the text at point [2] and start reading the novel for the second time, within the second loop. In this second reading, the signifier of the text is exactly the same. The letters are identical and the words are in the same order. The reading still starts with "Muchos años después, frente al pelotón de fusilamiento, el coronel Aureliano Buendía había de recordar aquella tarde remota en que su padre lo llevó a conocer el hielo" ["Many years later, as he faced the firing squad, Colonel Aureliano Buendía was to remember that distant afternoon when his father took him to discover ice"]. This second reading, however, is of a differently framed story because the first reading, by culminating in the convergence point of a mise en abyme, has rendered the second reading a fully embedded fabula. The second reading exists entirely within a frame—a frame that was inescapably assembled by the first reading.

The second batch of text that begins with "Muchos años después" ["Many years later"] conveys not just more but also different information. Once on the track of the second loop, the reader is reading not the cover-bound novel but rather Melquíades's parchments (which, the reader now knows, are essentially identical to the cover-bound novel). On the track of the second loop the reader is reading with Aureliano Babilonia's eyes, eyes which were only "created" by virtue of the first reading. The embedded fabula is the same as the frame fabula in terms of the text/signifier—the words that compose it are the same—but different in terms of context: The inner fabula is different because it is embedded within the outer fabula. It features a frame that the outer fabula does not. Likewise, the third batch of text—that on the next loop—is different from both the first and the second: The third is framed by two texts; the second is framed by one; the first is not framed.

Continuing with this construction, let's give the designation point [A] to an event in the novel—for instance, the insomnia plague. Point [B] stands for the onset of the insomnia plague *through the eyes of Aureliano Babilonia reading the parchments,* completely embedded within the primary fabula. Point [C] stands for the onset of the insomnia plague through the eyes of Aureliano Babilonia, reading the parchments through his own eyes' *first reading* of the parchments. This process continues down the abyss forever. It is because of this sense of a fathomless void in which each text contains in full another, necessarily "smaller" text that I suggested we picture a three-dimensional spiral, and not a two-dimensional spiral or a helix. The two-dimensional spiral lacks depth, and the helix consists of loops with the same radius. The three-dimensional spiral depicts that every successive reading is both situated on another plane and "smaller" than the one preceding it.

On the subject of forms that precisely depict the novel's mise en abyme, how about concentric circles? Concentric circles fall short of illustrating the recursion in *Cien años* for a simple but important reason: They do not indicate a connection between adjacent tiers such that one begets the next. By definition, each tier in mise en abyme depends on the one that precedes it. In concentric circles there's a sense of depth and there's also diminution, whereby each tier is smaller than the proceeding one, but there are no through lines that suggest that each tier emerges from the one that frames it. Through lines are a necessary component of a form that accurately portrays the narrative maneuver of recursions because they represent the passing through the point of convergence. We can see this passing through in the sentence I used some paragraphs ago to outline the infinite recursion in *Cien años*: Aureliano Babilonia is reading a narrative in which he reads about himself reading this narrative, in which he reads about himself reading this narrative, and so on. Each iteration of reading flows from one tier to the next—indeed *creates* the next, further embedded tier. Concentric circles do not illustrate this flow, this creation. They do not describe the particular relationship between tiers whereby each begets the next.

What about one, infinite loop, such as that set up in Joyce's *Finnegan's Wake*? The last line of the novel is "A way a lone a last a loved a long the." It is completed by the first line of the novel: "riverrun, past Eve and Adam's, from swerve of shore to bend of bay, brings us by a commodius vicus of recirculation back to Howth Castle and Environs." This circularity continues ad infinitum. Here, however, there is no embedding, so each iteration—each go around the loop—is identical to all the others. The existing information recurs, but no *new* information (other than that provided via readerly engagement) is added to the total information contained in and conveyed by the story. As such, circular recursion does not accomplish what embedded recursion does.

OTHER FORMS IN *CIEN AÑOS DE SOLEDAD*

We have considered the three-dimensional spiral, the two-dimensional spiral, the helix, concentric circles, and an ouroboroslike infinite loop. Only the first one truly fits mise en abyme. Let's take a look at other forms scholars have identified in *Cien años*. In "A Brief Story of Postmodern Plot," Catherine Burgass analyzes *Cien años* in terms of its circularity and treatment of time, particularly with regard to its ending: "The conclusion of the novel does not so much deconstruct the time of the story as bend it in a circle, effectively sealing the self-contained fictional world" (180). I agree that the novel's ending produces a sealed-off world, but, as I have argued, I believe the mise en abyme at the narrative's conclusion effectively casts the shape within which the self-contained fictional world is sealed into a regressive spiral.

Patricia Drechsel Tobin and Aníbal González employ the Moebius strip and Zeno's paradox, respectively, as models for *Cien años*. Tobin writes, "'A circle with a twist in it' may be a fair enough figure to describe this novel, whose narrative structure is aggressively linear, whose central theme is the circularity of time, and whose fictional universe is irrevocably cut off from our own. The figure is a Moebius strip" (176). Her study illustrates several ways in which this mathematical object can serve as a pertinent form in the analysis of *Cien años*. González, on the other hand, calls on Zeno's paradox to depict Aureliano Babilonia's reading of Melquíades's parchments:

> Translation is an endless activity, and appropriately, Aureliano, in his reading of the manuscript, is caught in a textual version of Zeno's paradox: he can approach the manuscript's end, but he can never reach or read it, because the text foretells his future and is always one step ahead of him; in order to reach the conclusion, he is forced, ineluctably, to cross a previous span of time (and of text) and no matter how much he skips, he will always have to read again and thus be chained to the text's tyrannical temporality. (75)

Tobin and González thus apply the Moebius strip and Zeno's paradox to their reading of the novel's last pages. The endless embeddedness produced at the point of convergence, however, inclines me toward the particular kind of circularity afforded by the three-dimensional spiral.

Mapping Entailments of Recursion

Reader Response Theory

We have identified a particular type of relationship between an actor (upon encountering a particular narrative maneuver) and the system upon which they act. Let's see what we can glean from an area of study dedicated to the adventures of this actor: reader response theory. In the foundational text *The Act of Reading,* Wolfgang Iser lays out the dialectic communication between text and reader. "The poles of text and reader, together with the interaction that occurs between them, form the ground-plan on which a theory of literary communication may be built" (ix). Many of Iser's insights on this dialectic relationship may apply to our present discussion in ways that he may not have intended but that inform and strengthen this chapter's argument nonetheless. For example, Iser writes, "Reading is an activity that is guided by the text; this must be processed by the reader, who is then, in turn, affected by what he has processed" (163). In our analogy this goes one step further—the dialectic communication continues: After the reader is affected by what they have processed at the point of convergence in the mise en abyme, the story itself—because it now requires substantial reconfiguration—is then affected by what the reader has processed, thereby continuing the "dialogue." Of course, as Iser points out, "with reading there is no *face-to-face situation.* A text cannot adapt itself to each reader with whom it comes in contact" (166). But it can and does demand reassessment as a consequence of the *act of reading,* in the ways I have outlined.

This dialectical, two-way process between text and reader is very much complicated by my assertion that sometimes—that is, when we have a perfect mise en abyme—the reading reshapes that which is read. The relationship is now not only a conversation. Rather, the reader reactualizes that which is read; by "conversing" with a story that contains infinite recursion, the reader reconsiders the story's storyness. This in effect buoys Iser's observation that "the text represents a potential effect that is realized in the reading process" (ix), which emerges as a metaphorical entailment: Quantum superposition also represents a particle's potential states, and those states transition into distinct classical alternatives at the moment of entanglement with the environment.

•

We have determined that a true mise en abyme is the narrative maneuver that best fits the present fi-sci analogy. In nonformulaic terms, we can say that the

TABLE 12. Recursion flowchart

	Phenomenon	Example	Expression
Science	1 Environmental entanglement	2 A quantum system interacts with a molecule.	3 [Environmental entanglement] disrupts the {phase coherence} of a <quantum system> such that the <quantum system> cannot maintain its (quantumness).
Template	—	—	4 [A mechanism] disrupts the {preservative} of an <entity> such that the <entity> cannot maintain its (unique properties).
Fiction	7 Mise en abyme	6 *Cien años de soledad,* Gabriel García Márquez	5 [Mise en abyme] disrupts the {structural and ontological boundaries} of a <story> such that the <story> cannot maintain its (storyness).

structural pattern of mise en abyme's impact can be mapped onto environmental entanglement's impact: In fiction, mise en abyme fundamentally affects the story; in quantum physics, environmental entanglement fundamentally affects the quantum system.

The Mise en Abyme || Environmental Entanglement fi-sci analogy (see table 12) renders a strange mechanism less strange. In this chapter we have seen how a certain narrative maneuver—the true mise en abyme—fundamentally affects its host story. Having located an avatar of the environmental entanglement pattern in fiction, we now have something with which we are already familiar that we can latch onto in order to conventionalize the peculiar science concept. Recognizing in fiction the shape of something that is challenging to conceptualize in science opens a portal to this phenomenon.[6]

6. Published with the permission of Cambridge Scholars Publishing.

CODA

Weave

The coda presents bow-tying contextualization in four moves. Here, we

1. complete figure 2, having gained the perspectives afforded by the six fi-sci analogy case studies;
2. set existing examples of (retconned) fi-sci pattern mapping into our analytical framework;
3. delimit the framework's constraints and capabilities, its boundaries and its potential;
4. situate this framework, in turn, among other frameworks that similarly seek the invariant.

1. Complete: What Kind of Fiction, Specifically

Figure 2 depicts conceptual category evolution plus science perceptual groups. The first step toward completing the figure is to update the examples of science—replacing the ones in the original figure with the examples of science featured in the six case studies. Before, under **Target [C1a] Straightforward science** were Newton's First Law of Motion and photosynthesis; under **Target [C1b] Complex science** were infrared light and evolution by natural selection; under **Target [C2] Counterintuitive science** were quantum entanglement and wave-particle duality. Now, leaving [C1a] blank for the moment since none of the science phenomena spotlighted in this book's case studies fall in the category of straightforward science, under **Target [C1b] Complex science** are genetic engineering (chapter 1), gametic reproduction (chapter 2), and clonal reproduction (chapter 3); under **Target [C2] Counterintuitive science** are spacetime curvature (chapter 4), quantum superposition (chapter 5), and environmental entanglement (chapter 6).

Next, let's add the fiction we mapped onto the science in these case studies—the sources we used to illuminate the targets. Since the science is listed in the target column—under the target constituent of **[C] Figurative analogy**—the fiction will appear in the source column, in new boxes directly under the source constituent of [C]. Thus, to the left of **Target [C1b] Complex science** we'll create the box **Source [C1b]**, and to the left of **Target [C2] Counterintuitive science** we'll create the box **Source [C2]**. Across from genetic engineering we'll list *Rayuela* and *Malas hierbas*; across from gametic reproduction, *Adios, Happy Homeland!*; across from clonal reproduction, works by Jorge Luis Borges. Across from spacetime curvature we'll list "Continuidad de los parques"; across from quantum superposition, the Muppets; across from environmental entanglement, *Cien años*.

Having mapped these seven examples of fictionality onto six examples of science, let us attempt to provide a more nuanced answer to the question, What kind of fiction? To do that, we'll ask another question: How can we differentiate between the kind of fiction listed in [C1b] and the kind of fiction listed in [C2]? In the introduction I suggested that the kind of fiction that facilitates our comprehension of scientific phenomena varies with the kind of science in question. Comparing three fictional islands—Robinson Crusoe's; Laputa; that in *Lost*—in table 2, I classified the first two as intuitive and the third as counterintuitive. I further differentiated between Crusoe's island and Laputa by classifying the former as possible and the latter as fantastical, impossible. Comparing table 2 to the fictions listed in [C1b] and [C2], we can say that the fictions in [C1b] are broadly intuitive, specifically, of the fantastical/impossible variety, and the fictions in [C2] contain salient elements that are counterintuitive. More than that: Now that we have analyzed these seven instances of fictionality in detail, we are in a position to render our differentiation more sophisticated.

To get there, I will once again borrow from Cohn, this time a series of adjectives she uses in "Metalepsis and Mise en Abyme." In this article Cohn makes two main distinctions: (1) between "metalepsis at the discourse level and metalepsis at the story level" (105); (2) between exterior metalepsis and interior metalepsis. It seems to me that "discourse level" lines up with "exterior," and "story level" lines up with "interior," so while Cohn does not explicitly state these alignments across the two distinctions, I proceed here with these alignments in mind and include what she writes about the first distinction in my analysis of the second distinction.

Discourse-level/exterior metalepsis "occurs between the extradiegetic and the diegetic level—that is to say, between the narrator's universe and that of his or her story" (106). This "results in a light-hearted and playful synchronization of the narration with the narrated events" (105) and is "relatively inoffensive" (106). On the other hand, story-level/interior metalepsis "occurs between two levels of the same story—that is to say, between a primary and secondary story, or between a secondary and tertiary story." This latter kind is "much more daring and shocking, also much more spectacular . . . a particularly troubling transgression. . . . Here, the boundary . . . is violated, leading to a confusion between distinct ontological levels" (106). Interior metalepsis "is an event that occurs in sudden and surprising isolation. . . . it stirs up in the reader a feeling of disarray, a kind of anxiety or vertigo" (110).

In my quest to differentiate between the fictions in [C1b] and the fictions in [C2], I gather what Cohn refers to as discourse-level/exterior metalepsis—playful, inoffensive iterations—under the umbrella of "metafiction." What Cohn calls story-level/interior metalepsis—iterations that are daring, shocking, and troubling, ones that may lead to feelings of disarray, anxiety, and vertigo—I collect under the umbrella of "metalepsis." I further subsume into my taxonomy the distinction Ryan makes between rhetorical and ontological metalepsis, explored in chapter 4. My umbrella of "metafiction" includes what Ryan terms "rhetorical metalepsis," and my umbrella of "metalepsis" includes what Ryan terms "ontological metalepsis." Rather than skate over nuances or propose a new system of distinctions, what I do with this two-umbrella paradigm is repurpose, or rearrange, the various distinctions others have made so that I can apply them to a distinction explored throughout this book: that between the intuitive and the counterintuitive.

The way I have described the behavior of intuitive but complex science systems—it is not beyond what we are capable of visualizing, but due to factors such as scale and scope these systems may pose more challenges to our understanding than ones that behave in clear, obvious ways—is, *mutatis mutandis,* the way I would describe metafiction. Foregrounding a blurring line between the diegetic and extradiegetic—more complex than straightforward, nonmeta fiction—the four examples of fiction in [C1b] all fit under the metafiction umbrella. The way I have described the behavior of counterintuitive science systems—as not corresponding to our embodied experience and observed reality, as tending toward incomprehensibility if not floating entirely beyond our notions of common sense—is the way I would describe metalepsis. Foregrounding a full-on breach of ontological strata, the three fictions in

[C2] fit under the umbrella of metalepsis.[1] It tracks that illuminating the more counterintuitive science requires the more jarring fiction.

Thus, with the information gathered from the six case studies, we can propose that complex science may be illuminated with metafiction, and counterintuitive science may be illuminated with metalepsis. We can now incorporate the two-umbrella paradigm terminology into the new boxes we'll add across from **Target [C1a] Complex science** and **Target [C2] Counterintuitive science**: to the left of the former, **Source [C1b] Metafiction**; to the left of the latter, **Source [C2] Metalepsis**. Note that the preceding analysis also helps us characterize **Source [C1a]**. If **Target [C1a]** is straightforward science, **Source [C1a]** is straightforward fiction, that is to say, the possible, realist kind, like Crusoe's island.

There are a couple of reasons the two-umbrella paradigm may not appear tidy, but these appearances are deceiving. First, one might point out that the [C1b] fictions I group under "metafiction" in fact feature moments of metalepsis. For instance, when listing metafictional elements in *Adios, Happy Homeland!* I mentioned that fictional character and purported editor Quain—based on a fictional character by nonfictional writer Borges—"collects" stories by authors that inhabit various ontological strata. One of them is a fictionalized version of the volume's actual author, Ana Menéndez. This is certainly a manifestation of "transgressions of narrative level," "produc[ing] a 'short-circuiting' of levels, calling [the] distinction [of levels] into question" (Pier 303). But I see it as a local example of metalepsis within a globally metafictional book. This is similar to how Nielsen et al. differentiate between local and global fictionality, whereby "global fictions can contain passages of nonfictionality, and global nonfictions can contain passages of fictionality" (67). It is also similar to how McHale shows, with reference to *Absalom, Absalom!*, that a narrative that "practices a poetics of the epistemological dominant" can contain a section—the novel's eighth chapter—that "dramatizes the shift of dominant . . . from an epistemological dominant to an *ontological one*" (10). As in these cases, the presence of moments of metalepsis in *Adios* does not take away from the book's global metafictional dominant.

Second, one might contend that if metalepsis is distinguished as troubling, inducing feelings of disarray and anxiety, the Muppets would not fit in the box. Indeed, audience reaction to the Muppets' metaleptic maneuvers is much more in line with the lighthearted, playful nature of inoffensive metafiction. I

1. I do not claim that all fantastical/impossible fiction is metafictional, nor that all counterintuitive fiction is metaleptic. Neither do I suggest that only metafiction can illuminate complex science, and only metalepsis can illuminate counterintuitive science. Solely based on our six case studies, however, these are generalizations we can glean.

would counter that the Muppets' ontological ambiguity *does* entail those more foreboding, jarring feelings, but these fade into the background because of the Muppets' irreverence and humor. In fact, *The Muppet Show*'s silliness and mayhem—its "overarching creative aesthetic" of nonsense (Abate 601)—was a direct response to the anxieties of its age. Abate offers the example of "Muppet Labs" skits featuring Dr. Bunsen Honeydew and his "hapless" assistant Beaker as taking place "against a backdrop of massive breakthroughs in medical procedure and scientific technology" (606). The skits reveal "the equal-parts societal fascination and cultural anxiety that surrounded the era's advancements in science and technology" (607). That the skits are playful and funny does not change the fact that they respond to feelings of uneasiness—different in nature from the ones metalepsis engenders but uneasiness, nonetheless. At its essence, the idea that a creature can be simultaneously fictional and nonfictional *is* unsettling; the notion that ontological boundaries are porous *is* disconcerting. Per Genette's interpretation, "the most troubling thing about metalepsis indeed lies in this unacceptable and insistent hypothesis, that the extradiegetic is perhaps always diegetic, and that the narrator and his narratees—you and I—perhaps belong to some narrative" (*Narrative Discourse* 236). But when that creature is Kermit the Frog, the levity we cannot help but associate with him washes out the uneasiness, installing playfulness at the forefront.

2. Set: Scientists Mapping Patterns from Fiction

Fi-sci pattern mapping—superimposing a phenomenon in fiction onto a similarly shaped phenomenon in science to render the latter more comprehensible, using fiction as a source to explicate the target of science—has actually been around for a while. "Fi-sci" and the demarcation system it entails are novel, but they describe a cognitive process that is already underway: Scientists have long been mapping patterns between fiction and science. To date I have encountered three established—that is, previously circulating—fi-sci analogies, each of which has been proposed by scientists to elucidate nature. In short, fictionality is already being wielded as a rhetorical instrument, as a means to the end of illuminating science. Since "fi-sci" is a neologism, none of these pairings was conceived as a fi-sci analogy. But in the spirit of Borges's "Kafka y sus precursores" ["Kafka and His Precursors"], I would like to show that we can retroactively consider them as such. To this end, I will derive a pattern-mapping flowchart from each analogy.[2]

2. I do not go into as much detail presenting these three examples as I did when analyzing the six case studies.

TABLE 13. Just-right flowchart

	Phenomenon	Example	Expression
Science	1 Circumstellar habitable zone	2 The conditions on planet Earth are just right for life (as we know it) to form.	3 Between [planetary temperatures that are too low] and {planetary temperatures that are too high} there exists a medium <range of temperatures> that is "just right" for (life).
Template	—	—	4 Between [one extreme] and {another extreme} there exists a medium <situation> that is "just right" for (a given entity).
Fiction	7 Optimal objects	6 "Goldilocks and the Three Bears"	5 Between [a chair; porridge; bed that is too large; cold; soft] and {a chair; porridge; bed that is too small; hot; stiff} there exists a medium <chair; porridge; bed> that is "just right" for (Goldilocks).

Optimal Objects || Circumstellar Habitable Zone

In the tale "Goldilocks and the Three Bears," Goldilocks tries three different versions of porridge, chair, and bed. In each triad, one version is too much of one extreme (cold porridge; large chair; soft bed), the second version is too much of the other extreme (hot porridge; small chair; stiff bed), and the third version is in between and "just right" (medium-temperature porridge; medium-size chair; medium-firm bed). Goldilocks thus comes to symbolize a search for optimal objects. The pattern "just right" has applications in numerous fields, including economics, machine learning, developmental psychology, and, perhaps most famously, astrobiology. Theoretical physicist Stephen Hawking explains, "Traditionally, given any star, scientists define the habitable zone as the narrow region around the star in which temperatures are such that liquid water can exist. The habitable zone is sometimes called the 'Goldilocks zone,' because the requirement that liquid water exist means that, like Goldilocks, the development of intelligent life requires that planetary temperatures be 'just right'" (152–53). Applying the same pattern to an even vaster set, cosmologist Paul Davies uses "the Goldilocks factor" to describe "the fitness of the universe for life" (xi): "Like the porridge in the tale of Goldilocks and the three bears, the universe seems to be 'just right' for life, in many intriguing ways" (3).

Whether in relation to circumstellar zones or to the whole cosmos, Hawking and Davies both leverage the similitude between a phenomenon in fiction and a phenomenon in science. Goldilocks's experience at the bears' house shares conceptual contours with the search for conditions hospitable to life, correlating the tale in the domain of fiction to astrobiology in the domain of science. In both cases, there exist circumstances whose attributes are "just right" for a given situation. Table 13 depicts the pattern-mapping flowchart I have derived from this analogy.

Object Evolution || RNA Evolution

A ship belonging to the mythological hero Theseus—perhaps best known for slaying the Minotaur—is featured in a famous thought experiment. Ship maintenance requires that every so often the ship's parts be replaced: one time the sails; another time the oars; then the masts, rudders, ropes. Eventually, every component of the ship will have been substituted with a newer version. The legendary question is this: If none of the original ship's parts remain, can we still consider it to be the same ship? There are variants of this paradox, including "Trigger's Broom" and "Grandfather's Axe"; in each case an item's components are replaced until it no longer comprises any of its original parts. The variations on this theme are different mappings of this pattern, but all are within the realm of fiction.

In a 2013 article about the origin of RNA, Nicholas V. Hud et al. map the "Ship of Theseus" pattern onto science (albeit referencing the "Grandfather's Axe" version thereof[3]):

> We have come to accept that life's informational polymers have changed in chemical structure since their emergence, which presents a quandary similar to the paradox of "My Grandfather's Axe." Here, we discuss reasons why all contemporary components of RNA—the nucleobases, ribose, and phosphate—are not likely the original components of the first informational polymer(s) of life. (466)

"As with the metaphorical grandfather's axe," they write, "we imagine that RNA evolution occurred in stages, with different chemical pieces being substituted multiple times and for different reasons" (472). Hud et al. propose that what happened to the axe over the course of its upkeep is similar to what happened over the course of proto-RNA's evolution into RNA. Though they

3. While Hud et al. use "Grandfather's Axe," I prefer the "Ship of Theseus" avatar for our purposes; the latter variation of the pattern is more apropos because in terms of RNA we are looking at the gradual substitution of *multiple* parts rather than the axe's *just two* parts.

TABLE 14. Gradual substitution flowchart

	Phenomenon	Example	Expression
Science	1 RNA evolution	2 Ribose sugar (a trifunctional connector [TC]), "was brought into contemporary RNA . . . through the replacement of ancestral...TCs" (Hud et al. 467).	3 Over time, all of [RNA's] original components have been replaced.
Template	—	—	4 Over time, all of [an entity's] original components have been replaced.
Fiction	7 Object evolution	6 "The Ship of Theseus" "Grandfather's Axe" "Trigger's Broom"	5 Over time, all of [the ship's] original components have been replaced.

do not use the terms *mapping* and *pattern,* Hud et al. explain how they map a pattern in one discipline onto another discipline:

> We liken RNA to the heirloom axe, where replacements of both the handle and the head of this metaphorical axe can also coincide with changes in function. Rather than accepting the notion that RNA is identical in chemical structure to the first informational polymer of life, we see persuasive reasons to place RNA as the penultimate member of a continuous series of polymers, with DNA being the most recent member of the series. (466)

Like Hawking and Davies do with Goldilocks's tale, Hud et al. take advantage of a congruence between a process in fiction and a process in science. The gradual replacement of the ship's parts—the object's evolution—shares a conceptual shape with the evolution from proto-RNA to contemporary RNA, correlating the story in the domain of fiction to biochemistry in the domain of science. In both cases, an entity's original components are eventually supplanted. Table 14 depicts the pattern-mapping flowchart derived from this analogy.

Authorial Directives || Epigenetic Marks

Mayevi Hadith Galindo Barrientos and Lesbia María Orellana Duarte begin their 2022 article on neuroembryology and neurodevelopment by defining epigenetics as "una ciencia nueva que estudia la información que se puede transmitir sin estar codificada en el ADN" ["a new science that studies the information that can be transmitted without being encoded in DNA"] (298 [Google Translate]). To clarify the concept, they offer several analogies to illustrate the rapport between epigenetics and genetics, the last of which would be right at home in the present volume's chapter on permutation:

> Para entender este término y su relación con la genética, se puede pensar como un libro donde la epigenética juega el papel de grapas que impiden que algunas páginas se lean o de resaltadores que resaltan algunas frases. También podría ser la tipografía, las letras mayúsculas, la cursiva, los signos de puntuación. *No cambia el contenido del libro, pero sí puede cambiar la manera en que esta se lee.* Podría hacerse esta analogía también como si fuera como el libro Rayuela de Julio Cortázar. Pueden existir dos libros iguales, la misma edición comprados el mismo día. Ambos libros serán como el material genético, pero la historia cambiará de acuerdo a si se lee de forma cronológica o con el instructivo del autor.
>
> [To understand this term and its relationship with genetics, you can think of it as a book where epigenetics plays the role of staples that prevent some pages from being read or highlighters that highlight some sentences. It could also be the typography, the capital letters, the italics, the punctuation marks. *It does not change the content of the book, but it can change the way it is read.* This analogy could also be made as if it were like the book *Hopscotch* by Julio Cortázar. There can be two identical books, the same edition bought on the same day. Both books will be like genetic material, but the story will change depending on whether it is read chronologically or with the author's instructions.] (299, emphasis added)

Each of the analogies here listed—staples, highlighters, typography, instructions about reading order—depicts something that does not change a text but does determine what is derived from this text. These descriptions all correspond with the "epigenetic view of differentiation," whereby "cells make a series of choices . . . that commit them to their eventual differentiated state. . . . The processes involve selective gene repression or derepression at an early stage, thereby restricting the range of final possible fates" ("Epigenetics"): The staples, highlighters, typography, and reading order will determine what will be read and/or how it will be read—they will determine the text's "final possible fate[ ]."

TABLE 15. Instructions flowchart

	Phenomenon	Example	Expression
Science	1 Epigenetic marks	2 A methyl group attached to a DNA element controlling the Agouti gene can turn transcription of the gene "off," thereby drastically impacting the phenotype of the mouse ("What Is Epigenetics").	3 [Epigenetic marks] lead to a {cell's} particular manifestation.
Template	—	—	4 [Instructions] lead to an {entity's} particular manifestation.
Fiction	7 Authorial directives	6 *Rayuela*, Julio Cortázar	5 [An author's directives] lead to a {story's} particular manifestation.

What is common to the analogies Galindo Barrientos and Orellana Duarte list is their relation to aspects of textuality (pages; sentences; typography; punctuation; books). We could therefore understand this set of analogies as an entailment of the ur-analogy between text and genome described by Lévi-Strauss and Mukherjee and discussed in this book's introduction. But while these analogies all pertain to textuality, only the last one mentioned above—the one about *Rayuela*—pertains to a *work of fiction*. Table 15 depicts the pattern-mapping flowchart derived from this final analogy.

•

We have just applied the analytical steps taken in each of this book's case studies to three instances of pattern mapping across fiction and science. In that they all illuminate science with recourse to fictionality—in other words, use fictionality as a means to the end of better understanding science—Optimal Objects || Circumstellar Habitable Zone, Object Evolution || RNA Evolution, and Authorial Directives || Epigenetic Marks are fi-sci analogies. Let's include them in figure 4, which is an updated version of figure 2 that also contains all the additions I described in section "Complete," above.

It is interesting to notice that what we find here is in line with the rest of this book. Of the three newly labeled fi-sci analogies, the first two feature a straightforward science concept; the third features a science concept that is more complex. Correspondingly, the first two feature straightforward fiction;

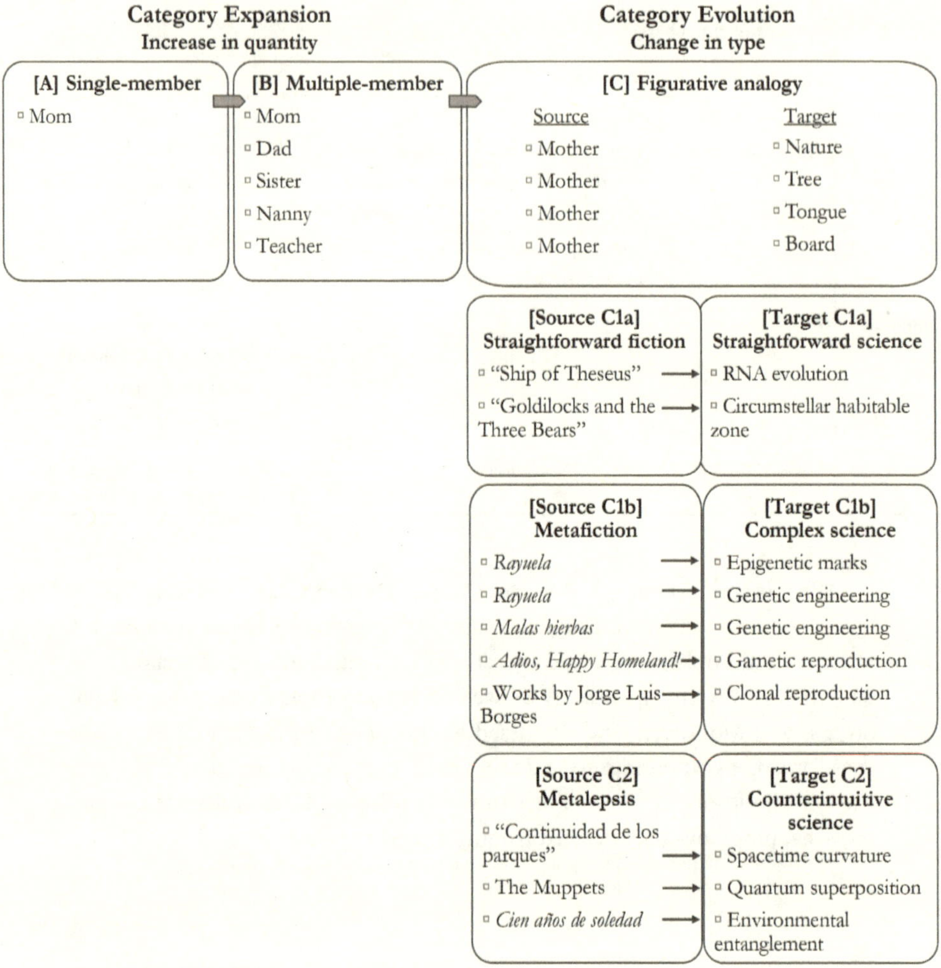

FIGURE 4. Conceptual category evolution, updated.

the third features metafiction. The patterns of the first two duos—“just right” and “gradual substitution”—do not make recourse to metafiction. In the fiction avatars of these patterns, there are no blurring lines between diegetic and extradiegetic levels. If three talking humans were to stand in for the three talking bears—which is acceptable since the “just right” pattern does not hinge on the three characters being bears, so substituting them with humans does not affect the pattern—Goldilocks's story would be straightforward fiction. And the story of Theseus's ship does not even need tinkering—it is straightforward fiction from the get-go. Straightforward fiction is not the kind called for by **Target [C1b]** and **Target [C2]**, which list science phenomena whose conceptualization expressly benefits from fictionality of the meta variety. All six original case-study patterns pivot on something meta—whether metafiction (complex) or metalepsis (counterintuitive). That is also the case with the Authorial Directives || Epigenetic Marks analogy presented by Galindo Barrientos and Orellana Duarte. It involves, as they say, the "author's instructions" and is therefore by definition self-referential, reaching beyond the boundaries of the text.

3. Delimit: What It Is; What It Isn't

Let's calibrate the fi-sci model based on the preceding analysis. What follows are two considerations meant to help adjust the implications and expectations of the pattern-mapping analytical framework.

3.1. Analogy ≠ Identicality

When we compare a phenomenon in fiction to a phenomenon in science we are making an analogy, saying "this thing is like that thing." *Like*, not *is*. The map is not a one-to-one plotting across every aspect of the source and the target. If it were, such correlation would cease to be an analogy; the concepts under comparison would be identical. And so, it is important to bear in mind two caveats of the fi-sci pattern-mapping framework: (1) Not all nodes will map. As I've explored in each chapter's section on metaphorical entailments, in every analogy inevitably some nodes will have analogs—otherwise no comparison could be made; others will not—if they all mapped we'd have an identicality. Recall Cajetan's analogy of proportionality discussed in chapter 3—A:B :: C:D, whereby A is to B as C is to D—and consider the part-to-whole analogy Petal : Flower :: Wheel : Car ("Verbal Analogies"). A petal is like a wheel in that both form parts of a whole of a flower and a car, respectively. But there are many ways in which a petal is not like a wheel. In fact, I see the two as being different in more ways than they are similar. This is a good example of two concepts with nodes that have analogs and nodes that do not. When presenting fi-sci analogies in the classroom, instructors might begin by distinguishing "[*x*] is [*y*]" from "some qualities of [*x*] are analogous to some qualities of [*y*]."

Speaking of using this pattern-mapping framework in the classroom, I submit it would be exceedingly rare for an instructor teaching a fi-sci analogy to hold expertise in both domains. Given academe's desideratum for specialization, most of the time those who concoct and/or convey a fi-sci analogy will know more about one area than the other. When I, as a literature professor, show my students the Ambiguity flowchart, I am by no means even attempting to teach them about quantum superposition (that would be more absurd than a cat that is simultaneously dead and alive); when my particle physicist colleague does the same, they are by no means even attempting to teach their students narrative theory. Hence, caveat (2): Having made the analogy between [*x*] and [*y*], students have not learned everything there is to know—that is, all that they would have been taught in a lesson specifically about one or the other in a class devoted to that subject—about either. Instead, they have come to see how an aspect of [*x*] is analogous to an aspect of [*y*], meaning they have developed integrative thinking, an expanded perspective on whichever expertise they're cultivating in the course of their studies. In short, these students are gaining *broad*, cross-disciplinary proficiency. *In-depth* subject proficiency in [*x*] or [*y*] is to be gained in a different lesson.

3.2. Mutual Illumination

I began this project with the goal of illuminating science with fiction. Since my expertise is in the latter, it constitutes my source domain; the former—that which, for me, requires illumination, constitutes the target domain. Elena Semino explains, "Cognitive metaphor theorists emphasize that target domains typically correspond to areas of experience that are relatively abstract, complex, unfamiliar, subjective or poorly delineated. . . . In contrast, source domains typically correspond to concrete, simple, familiar, physical and well-delineated experiences" (6). As someone trained in narrative theory sans an academic background in science, I see the fi of fi-sci as familiar and the sci as unfamiliar. However, there is no reason to discount the possibility that the analogy could work the other way around. It is conceivable that someone more familiar with the science would use their understanding of *it* to better comprehend what is going on in the similarly patterned fiction, such that the source and target would effectively be reversed. Fi-sci might thus function in multiple directions: For some people and on some occasions the fiction will shed light on the science; for some people and on some occasions the science will shed light on the fiction; for some people and on some occasions the two will be mutually illuminating.

Black's interaction view of metaphor foregrounds this principle of a metaphor's two components acting upon each other. Using as an example the statement "man is a wolf," he writes, "If to call a man a wolf is to put him in a special light, we must not forget that the metaphor makes the wolf seem more human than he otherwise would" (44). Thus it is not just the source that has an impact on our understanding of the target, but also the target that has an impact on our understanding of the source. As Ken Baake puts it, "For Black and other interactionists, the metaphoric process moves in two directions, which simply means that the tenor and vehicle affect each other" (63).[4] Citing Richards's portrayal of "a reader forced to 'connect' the two ideas," Black makes a powerful statement: "In this 'connection' resides the secret and the mystery of metaphor" (Richards 125; Black 39). Black offers notable alternatives to describing "the 'interaction' of two thoughts 'active together'": "their 'interillumination' or 'co-operation'" (39). With its own nuances, each of these descriptors conveys the notion of mutual benefit.

In *Narrating the Mesh,* Caracciolo makes a similar observation. Metaphors, he argues, "generate semantic and affective 'feedback' via their mapping of source and target domains. Metaphorical language has the potential for triggering far more than a linear, unidirectional transfer of meanings from source to target; it can offer new insights into *both* the source and the target domain" (162). The addition of "feedback" to Max Black's collection "interaction," "interillumination," and "co-operation" makes even more palpable analogies' bidirectionality.

Daniel Aureliano Newman provides another perspective on the bidirectionality of analogy, discussing the fi-sci framework's potential in pedagogy. "For literary scholars," Newman writes, "fi-sci pattern mapping offers interpretive possibilities beyond teaching science; it can form the basis of a hermeneutic process in which scientific and literary knowledge inform each other recursively" (312). Having published extensively on intersections of narrative and science, Newman sees it like this:

> Science and literature are thoroughly interwoven, mutually informing each other: scientific models inspire literary innovations while fictional narratives, in the form of thought experiments, propel scientific theorizing. In this conception, a narrative's fictionality serves to lower barriers to scientific learning, even as the scientific models or principles in question spur new ways of analyzing and understanding, as well as constructing, fictional forms. . . . As I use it, then, the hyphen in "fi-sci" indicates not unidirectional transfer from fictional source to scientific target but also the potential for productive multidirectional exchange. (312)

4. *Tenor* and *vehicle* are roughly interchangeable with *target* and *source,* respectively.

Thus mutually enlightening, perhaps when brought together in analogy the fi and the sci enable understanding that is greater than each can provide on its own. This recalls Baake's remarks about "theorists in the constructivist camp." For them, "metaphoric statements involve what Richards refers to as the 'interanimation' between words (69), or what Black refers to as an 'interaction' ("More" 27). In both cases, the meaning of a statement is *greater . . . than the sum of its individual lexical components*" (62, emphasis added).[5]

Newman agrees with my thesis that, as he puts it, "one of the affordances of fictionality is to lower barriers to scientific comprehensions." But he adds, "Reciprocally . . . fiction that borrows from and even merely names scientific theories, models and phenomena avails itself of new kinds of reading, which might not otherwise be admissible" (313). I can support this assertion with a callback to the duet of chapters 2 and 3. In chapter 3's final section, I noted that the benefit of focusing on the multiplicity characteristic of proliferation is that it allows us to gain a new insight into the maneuvers and devices of fiction. Specifically, without fi-sci pattern mapping we may not have been prompted to detect a relationship between transtextuality and authorial motifs. Having used fi-sci pattern mapping, we recognize that they correspond in that both are mechanisms by which literary beings (texts) proliferate, and they differ in how multiplicity manifests in each. Exploring transtextuality and authorial motifs with multiplicity in mind allows us to discern this particular relationship between them. In this way, we use our understanding of the fi to illuminate the sci, and then turn our understanding of the sci back to further enhance our understanding of the fi. So while the crux of my project is to establish that fictionality has the rhetorical power to promote science understanding, it is certainly worth noting that illumination can result from traversing the analogy in more than one direction.

4. Situate: "The Quest for the Invariant"—And Its Variations

> "Why is a raven like a writing desk?"
>
> —Lewis Carroll, *Alice's Adventures in Wonderland*

As we have seen, fi-sci pattern mapping is a familiar friend. Its particular wielding of fiction as a source to illuminate the target of science is not new.

5. Baake continues, "Richards conveys this notion of a greater sum when he writes that, 'a word is always a cooperative member of an organism' (69)" (62).

Interestingly, when we zoom out and consider cross-disciplinary pattern mapping in a macro sense, we find that this is not new either. Reflect on Bronowski's account of Johannes Kepler's method:

> Kepler's mind, we know, was filled with . . . fanciful analogies. . . . Kepler wanted to relate the speeds of the planets to the musical intervals. He tried to fit the five regular solids into their orbits. None of these likenesses worked, and they have been forgotten; yet they have been and they remain the stepping stones of every creative mind. Kepler felt for his laws by way of metaphors, he searched mystically for likenesses with what he knew in every strange corner of nature. (12)

In the introduction, I quoted Bronowski's view that "all science is the search for unity in hidden likenesses" (13). He offers the examples of Ernest Rutherford and Niels Bohr, who likened the atom to the solar system (12); Hideki Yukawa, who likened the forces that hold together an atom's nucleus to solid pellets; Michael Faraday, who likened electricity to magnetism (13); Clark Maxwell, who likened electricity and magnetism to light; Albert Einstein, who likened time with space, and mass with energy (15); Isaac Newton, who likened an apple to the moon. "Seiz[ing] a likeness between two unlike appearances. . . . Newton traced in them two expressions of a single concept, gravitation" (14). Newton mapped the pattern of gravity working on the apple onto the pattern of gravity working on the moon. In short, "The progress of science is the discovery at each step of a new order which gives unity to what had long seemed unlike" (15). A very similar notion is reflected in this question presented by Kenneth Burke: "As the documents of science pile up, are we not coming to see that whole works of scientific research, even entire *schools,* are hardly more than the patient repetition, in all its ramifications, of a fertile metaphor?" (95).

We see the same thing when we think about art. "Science is nothing else than the search to discover unity in the wild variety of nature—or more exactly, in the variety of our experience. Poetry, painting, the arts are the same search, in Coleridge's phrase, for unity in variety" (Bronowski 16).[6] Analyzing examples from William Shakespeare, Christopher Marlowe, Alexander Pope, Bolingbroke, Johann Wolfgang von Goethe, and W. B. Yeats, Bronowski

6. Bronowski is here referring to something he mentions just before the sentence I quoted: "When Coleridge tried to define beauty, he returned always to one deep thought: beauty, he said, is 'unity in variety'" (16). Bronowski appends a note in which he tracks Coleridge's statement further back; I reproduce the note here to demonstrate the numerous iterations of this notion: "In one of the places in which Coleridge put forward this definition . . . he traced it back to Pythagoras: 'The safest definition, then, of Beauty, as well as the oldest, is that of Pythagoras: The reduction of many to one'" (22n8).

concludes, "The discoveries of science, the works of art are explorations—more, are explosions, of a hidden likeness" (19).

Let us back up to the micro of fi-sci again for a moment and think about this: What is the impetus for anyone to locate similarities between two very different fields? Lévi-Strauss's mother told him about an insight he had at two years of age. He recounts,

> Still unable to read, of course, I claimed that actually I was able to read. And when I was asked why, I said that when I looked at the signboard on shops—for instance, *boulanger* (baker) or *boucher* (butcher)—I was able to read something because what was obviously similar, from a graphic point of view, in the writing could not mean anything other than "bou," the same first syllable of *boucher* and *boulanger.* Probably there is nothing more than that in the structuralist approach; it is the quest for the invariant, or for the invariant elements among superficial differences. (8)

The quest for the invariant is our human propensity to look for connections, to reason by locating similarities among things, "to find what is common to them all" (Lévi-Strauss 9). One path in this quest for the invariant is described in this book's fi-sci pattern-mapping approach. But there are others. In chapter 3 I proposed that Borges's iterations are *about* iterations. He sums up this idea, as I've noted, by suggesting that "perhaps universal history is the history of the various intonations of a few metaphors." It seems fitting that if there are a handful of metaphors/patterns, if "nature has only a limited number of procedures at her disposal" (Lévi-Strauss 10), then similarly, there should exist a handful of ways to approach the quest for the invariant. The present move of the coda showcases variations on the quest for the invariant. Or, to make it meta and perhaps convoluted, it is the result of my quest to find the invariant in a handful of approaches to the quest for the invariant. Think of this in terms of Plato's ideal forms: There are patterns out there, and they manifest in different avatars in different fields. What follows is an overview of five avatars of *that* idea.

Isomorphism

Hofstadter first defines isomorphism with relation to the canons of Johann Sebastian Bach: "The idea of a canon is that one single theme is played against itself. This is done by having 'copies' of the theme played by the various participating voices." Canons may escalate in complexity, such as "when the 'copies' of the theme are staggered not only in *time*, but also in *pitch*"; another "stage of complexity in canon construction is to *invert* the theme" (8). So canons vary in complexity, but what they have in common is that "every type of 'copy' preserves all the information in the original theme, in the sense that the theme is fully recoverable from any of the copies. Such an information-preserving transformation is often called an *isomorphism*" (9). The information-preserving transformation—an "underlying idea" (11)—is what we have been referring to as a pattern. The particular pattern Hofstadter traces through *Gödel, Escher, Bach* is one that he calls a "Strange Loop," which "occurs whenever, by moving upwards (or downwards) through the levels of some hierarchical system, we unexpectedly find ourselves back where we started" (10).

Here Hofstadter introduces the book's first instance of transdisciplinary/transmodal strange loop isomorphism: "Look, for example, at [M. C. Escher's] lithograph *Waterfall*, and compare its six-step endlessly falling loop with the six-step endlessly rising loop of [Bach's] 'Canon per Tonos.' The similarity of vision is remarkable. Bach and Escher are playing one single theme in two different 'keys': music and art" (11–12). Soon Hofstadter extends the parallel to include "a mathematical counterpart" (15)—Kurt Gödel's incompleteness theorem—thus completing the trifecta of his book's title: "Just as the Bach and Escher loops appeal to very simple and ancient intuitions—a musical scale, a staircase—so this discovery, by K. Gödel, of a Strange Loop in mathematical systems has its origins in simple and ancient intuitions" (15, 17).

In a later chapter, Hofstadter expands the definition of "isomorphism," noting that it "applies when two complex structures can be mapped onto each other, in such a way that to each part of one structure there is a corresponding part in the other structure, where 'corresponding' means that the two parts play similar roles in their respective structures" (49). Using Hofstadter's terminology we can say that fi-sci analogies are specific isomorphisms—those that focus on correspondences between fictionality and science. One difference between my formulation of fi-sci and Hofstadter's formulation of isomorphism is that in the former each part of a structure in one domain does not always have a counterpart in the other domain—there are nodes that map, and nodes that do not. As such, isomorphisms may be seen as the exception that proves the rule.

Boundary Objects

Susan Leigh Star and James R. Griesemer define boundary objects as "objects which are both plastic enough to adapt to local needs and the constraints of the several parties employing them, yet robust enough to maintain a common identity across sites. . . . They have different meanings in different social worlds but their structure is common enough to more than one world to make them recognizable, a means of translation" (393). Introduced in an article about collaboration among participants at Berkeley's Museum of Vertebrate Zoology, boundary objects enable communication and translation. The article "takes as its starting point the acknowledgement that scientific work is heterogenous and requires cooperation between multiple actors" (Stoytcheva). Star and Griesemer propose that an activity "central for translating between viewpoints" and making such cooperation possible is the development of boundary objects (387).

With the analytic concept of boundary objects, Star and Griesemer introduce a more "egalitarian mode of intergroup communication" (Stoytcheva); indeed, the article wherein the concept originates appears in the journal *Social Studies of Science*. But there is no reason the notion of boundary objects cannot itself serve as a boundary object, plastic enough to have applications in varied milieus. I submit that we can understand fi-sci's patterns as boundary objects. A given pattern's avatar in fiction and its avatar in science are manifestations of the object's plasticity (it "adapt[s] to local [disciplinary] needs and constraints"); that the same pattern is at play in both domains shows that it "maintains a common identity across sites."

Biomimicry

In the subtitle of her seminal study *Biomimicry*, Janine M. Benyus refers to it as "innovation inspired by nature." It is in the name: Biomimicry involves copying nature. Biomimics look to see how nature "does things" and apply it to human innovation. A prominent, relatively recent example of biomimicry is the redesign of Japan's Shinkansen bullet train. This train was superfast, but it caused a significant disturbance: Whenever it exited a tunnel it would create a sonic boom that could be heard hundreds of meters away. "An engineering team was brought in to design a quieter, faster, and more efficient train" ("The World Is" 00:31–00:36). Their models? Birds. The pantograph was inspired by an owl's feathers; the pantograph's supporting shaft was inspired by the Adelie penguin's belly; the train nose was inspired by the Kingfisher's beak. The redesigned train was "10% faster, used 15% less electricity, and stayed under the 70 dB noise limit in residential areas" (1:50–1:57). Essentially, the engineers mimicked birds to create a better train. As James Geary puts it, the biomimic's motto is "ask not how I can solve that problem but how nature has already solved it" (206).

Biomimicry was published in 1997, but the practice has been around for centuries. "Emulating nature for designs that work is not new, and biomimicry (or biomimetics) is a (re)emerging discipline of ancient practice found within the majority of indigenous cultures" (Daynes). Alistair Daynes provides a timeline of biomimicry, offering as the earliest example the case of Lu Ban (c. 507–444 BCE), who is "credited with inventing the umbrella, inspired by a giant lily leaf; and the saw, inspired by a serrated vine stem." A much more recent example is Velcro, invented in 1941 by George de Mestral, who was inspired by the hooked spines of the cocklebur plant to create the artificial hook-and-loop mechanism (Daynes).

Geary writes, "Biomimics interrogate flora and fauna to find models that can be carried over into the human environment" (206); this carrying over is pattern mapping. Biomimics map a pattern they perceive in nature onto something that is made by humans, and the benefits entailing the design of the former become the benefits that entail the latter. Biomimicry differs from fi-sci pattern mapping in terms of the wielder's goal: The pattern mapper recognizes and locates similarities between entities in different domains for the purpose of mutual illumination, while the biomimic applies the pattern at play in one entity in order to create or improve an entity in another domain. Still, we can think of fi-sci pattern mapping as biomimicry plus fiction: Both bioinspired design and fi-sci leverage analogical reasoning to solve problems across disciplines.

Superconcepts

Per the National Academy of Sciences, interdisciplinary research "integrates information, data, techniques, tools, perspectives, concepts or theories from two or more disciplines or bodies of specialized knowledge" (26). This is the system- and problem-based approach to interdisciplinarity (Gombrich; Wilson, *Knowledge*), also described as the integrative-synthesis model (Barry et al.). There is another approach, called concept-based interdisciplinarity, "which comes about through the *transferability of concepts*" (Wilson, *Knowledge* 35). This latter approach turns on the notion of a *superconcept*, defined as a concept that "has usually been established in a particular discipline but turns out to have a much wider range of applications across many disciplines" (Wilson, "Routes"). A quintessential example of a superconcept is evolution:

It "arose in biology," but soon "developed into a concept of great explanatory power in fields as diverse as computer science (genetic algorithms) and cultural studies (memes)" (Gombrich 346). Similarly to how "just right" conditions are at play both in Goldilocks's tale and in astrobiology, the evolution superconcept manifests in disparate domains. It is this second approach to interdisciplinarity that characterizes fi-sci pattern mapping; superconcepts are patterns. Thus, evolution is a pattern with avatars arising in biology, computer science, and cultural studies.

Meta-Concepts

Trained as a particle physicist, Vandana Singh is a professor of physics and the environment, as well as an award-winning author of science fiction. In 2024 she published *Teaching Climate Change: Science, Stories, Justice,* in which she sets out her transdisciplinary approach to climate science pedagogy. At its core is a conceptual structure "for making sense of climate change" (89): three ideas Singh calls "meta-concepts," that "together form a conceptual scaffolding that enables a deeper and wider understanding of the basic science of the climate problem" (53).

a. Balance/imbalance

For instance, "an imbalance in the carbon cycle resulting in increasing carbon dioxide levels in the atmosphere will, through the greenhouse effect, cause an imbalance in Earth's energy budget" (57).

b. Critical thresholds

"Ecosystems, for example, can go from states of relative stability to abrupt change that is irreversible over those timescales" (62).

c. Complex interconnections

An example of a complex system is that "small changes in insolation due to orbital variations are connected with changes in ocean currents and atmospheric CO_2 levels that together give rise to the dramatic phenomenon of glacial-interglacial cycles" (68).

Each of these interconnected meta-concepts thus describes an aspect of the climate crisis. But these ideas are "broad enough that they have meaning in other disciplines" and as such are transdisciplinary (73). Indeed: All three concepts introduced here in the context of climate science manifest in economics, politics, and psychology. In this way the meta-concepts are, in the parlance of the present study's analytical framework, patterns, with avatars arising across disciplines. Singh in fact sees these meta-concepts as boundary objects, "since they have multiple interpretations that become specific in particular contexts" (123). She explains why she considers stories to be boundary objects as well: "Viewed from different disciplinary perspectives, they acquire specific, sharper meanings" (90).

•

These five variations of the quest for the invariant, presented here chronologically, are not the only endeavors of this kind. All the way back in the introduction, for example, we explored Bejan's constructal law of physics, whereby "design in nature"—the relationship between form and function—results from the flow of currents, be they water, electricity, traffic, or social organizations. We can now situate fi-sci pattern mapping among other systems, all designed to find the invariant across disciplines. This list of five variations—isomorphism; boundary objects; biomimicry; superconcepts; meta-concepts—is itself a mapping of a pattern: the pattern of pattern mapping.

ACKNOWLEDGMENTS

This book is motivated by the power of analogy. I hereby reach for an analogy one more time to acknowledge the colleagues, friends, and family who shaped this book's ideas and fine-tuned their presentations. For this last analogy I go full circle, drawing inspiration from Bejan's constructal law, first broached in the introduction.

Bejan and Zane write,

> Everything that moves, whether animate or inanimate, is a flow system. All flow systems generate shape and structure in time in order to facilitate this movement across a landscape filled with resistance (for example, friction). The designs we see in nature are not the result of chance. They arise naturally, spontaneously, because they enhance access to flow in time. (3)

If this book is a flow system, its medium is narratology, specifically, the rhetorical approach to fictionality. And the only reason this book's ideas resemble a system is that James Phelan identified that particular methodology as the flow's conduit. Beforehand, all I had were a handful of quirky connections between science I find fascinating and stories I love and little notion as to what's at stake. In response to my agonizing about the "so what" of it all, Jim revealed that it was rhetoric: Fictionality illuminating science is fictionality getting something done. Jim's key insight is that rhetorical theorizing about fictionality is the channel through which fi-sci pattern mapping flows.

APPENDIX

Variations on a Theme

In what follows, all quotations in the original Spanish appear in Borges's *Obras completas, 1923–1972* and *Obras completas, 1975–1985*. All quotations in English appear in Borges's *Collected Fictions,* with the exception of three texts: "Kafka and His Precursors" and "John Wilkins' Analytical Language" appear in Borges's *Selected Non-Fictions*; "Partial Magic in the *Quixote*" appears in *Labyrinths.* In these I will cite the translator. All emphases are original.

Variations

I. Singular

- (Given eternity) One object is all objects; one man is all men; all men are one man
  1. "Tlön, Uqbar, Orbis Tertius"
     i. "Todos los hombres, en el vertiginoso instante del coito, son el mismo hombre. Todos los hombres que repiten una línea de Shakespeare, *son* William Shakespeare" ["All men, in the dizzying instant of copulation, are the same man. All men who speak a line of Shakespeare *are* William Shakespeare"] (*Obras completas, 1923–1972* 438n1 [Hurley 76n4]).
  2. "La forma de la espada" ["The Shape of the Sword"]

i. "Lo que hace un hombre es como si lo hicieran todos los hombres. . . . Acaso Schopenhauer tiene razón: yo soy los otros, cualquier hombre es todos los hombres" ["Whatsoever one man does, it is as though all men did it. . . . Schopenhauer may have been right—I am other men, any man is all men"] (*Obras completas, 1923–1972* 493, 494 [Hurley 141]).

3. "El inmortal" ["The Immortal"]
   i. "Nadie es alguien, un solo hombre inmortal es todos los hombres" ["No one is someone; a single immortal man is all men"] (*Obras completas, 1923–1972* 541 [Hurley 191]).
4. "Los teólogos" ["The Theologians"]
   i. "El tratado era límpido, universal; no parecía redactado por una persona concreta, sino por cualquier hombre o, quizá, por todos los hombres" ["John's treatise was limpid, universal; it seemed written not by a particular person, but by any man—or perhaps all men"] (*Obras completas, 1923–1972* 552 [Hurley 203]).
5. "*Deutsches Requiem*"
   i. "es natural que piense en mis mayores, ya que tan cerca estoy de su sombra, ya que de algún modo soy ellos" ["it is natural that I should think of my elders, since I am come so near their shadow—since, somehow, I am they"] (*Obras completas, 1923–1972* 576 [Hurley 229]).
6. "El Zahir" ["The Zahir"]
   i. "no hay moneda que no sea símbolo de las monedas que sin fin resplandecen en la historia y la fábula" ["there is no coin that is not the symbol of all the coins that shine endlessly down throughout history and fable"] (*Obras completas, 1923–1972* 590–91 [Hurley 244]).
7. "El Aleph" ["The Aleph"]
   i. "un pájaro que de algún modo es todos los pájaros" ["a bird that somehow is all birds"] (*Obras completas, 1923–1972* 625 [Hurley 282]).
8. "*Everything and Nothing*"
   i. "Nadie fue tantos hombres como aquel hombre" ["No one was as many men as that man"] (*Obras completas, 1923–1972* 803 [Hurley 320]).

   9. "El Congreso" ["The Congress"]
      i. "un pájaro que es todos los pájaros, un sol que es todas las estrellas y el sol" ["a bird that is all birds, a sun that is the sun and yet all stars"] (*Obras completas, 1975–1985* 31 [Hurley 435]).
- One object or symbol encompasses everything, or stands for God, or the word of God, or contains the totality of all things
   1. "La biblioteca de Babel" ["The Library of Babel"]
      i. "debe existir un libro que sea la cifra y el compendio perfecto *de todos los demás*" ["there must exist a book that is the cipher and perfect compendium of *all other books*"] (*Obras completas, 1923–1972* 469 [Hurley 116]).
   2. "El milagro secreto" ["The Secret Miracle"]
      i. "El bibliotecario le dijo: *Dios está en una de las letras de una de las páginas de uno de los cuatrocientos mil tomos del Clementinum*. . . . Vio un mapa de la India, vertiginoso. Bruscamente seguro, tocó una de las mínimas letras. Una voz ubicua le dijo: *El tiempo de tu labor ha sido otorgado*" ["*God,* the librarian said, *is in one of the letters on one of the pages of one of the four hundred thousand volumes in the Clementine*. . . . He saw a map of India—a dizzying page. Suddenly certain, he touched one of the tiny letters. A voice that was everywhere spoke to him: *The time of your labor has been granted*"] (*Obras completas, 1923–1972* 511 [Hurley 160]).
   3. "El Zahir" ["The Zahir"]
      i. "Tal vez quiso decir que no hay hecho, por humilde que sea, que no implique la historia universal y su infinita concatenación de efectors y causas. Tal vez quiso decir que elm undo visible se da entero en cada representación. . . . Los cabalistas entendieron que el hombre es un microcosmo, un simbólico espejo del universo . . ." ["Perhaps he was trying to say that there is nothing, however humble, that does not imply the history of the world and its infinite concatenation of causes and effects. Perhaps he was trying to say that the visible world can be seen entire in every image. . . . The Kabbalists believed that man is a microcosm, a symbolic mirror of the universe . . ." (248)] (*Obras completas, 1923–1972* 594, 595 [Hurley 248]).

ii. ". . . quizá detrás de la moneda esté Dios" [". . . perhaps behind the coin is God"] (*Obras completas, 1923–1972* 595 [Hurley 249]).

4. "La escritura del Dios" ["The Writing of the God"]

i. "Una montaña podía ser la palabra del dios, o un río o el imperio o la configuración de los astros" ["A mountain might be the word of the god, or a river or the empire or the arrangement of the stars"] (*Obras completas, 1923–1972* 597 [Hurley 251]).

ii. ". . . aun en los lenguajes humanos no hay proposición que no implique el universo entero . . ." [". . . even in the languages of humans there is no proposition that does not imply the entire universe . . ."] (*Obras completas, 1923–1972* 597 [Hurley 252]).

iii. "Un dios, reflexioné, sólo debe decir una palabra y en esa palabra la plenitud. Ninguna voz articulada por él puede ser inferior al universo o menos que la suma del tiempo. Sombras o simulacros de esa voz que equivale a un lenguaje y a cuanto puede comprender un lenguaje son las ambiciosas y pobres voces humanas, *todo, mundo, universo*" ["A god, I reflected, must speak but a single word, and in that word there must be *absolute plenitude.* No word uttered by a god could be less than the universe, or briefer than the sum of time. The ambitions and poverty of human words—*all, world, universe*—are but shadows or simulacra of that Word which is the equivalent of a language and all that can be comprehended within a language"] (*Obras completas, 1923–1972* 598 [Hurley 252]).

iv. ". . . hay quien lo hay visto a Dios en un resplandor, hay quien lo ha percibido en una espada o en los círculos de una rosa. Yo vi una Rueda altísima . . . Entretejidas, la formaban todas las cosas que serán, que son y que fueron . . . Vi el universo y vi los íntimos designios del universo" [". . . one man has seen God in a blinding light, another has perceived Him in a sword or in the circles of the rose. I saw a Wheel of enormous height. . . . It was made of all things that shall be, that are, and that have been, all intertwined. . . . I saw the universe and saw its secret designs"] (*Obras completas, 1923–1972* 598, 599 [Hurley 253]).

5. "El Aleph" ["The Aleph"]
   i. "El diámetro del Aleph sería de dos o tres centímetros, pero el espacio cósmica estaba ahí . . ." ["The Aleph was probably two or three centimeters in diameter, but universal space was contained inside it . . ."] (*Obras completas, 1923–1972* 625 [Hurley 283]).
   ii. ". . . el universo está en el interior de una de las columnas de piedra que rodean el patio central . . ." [". . . the universe lies inside one of the stone columns that surround the central courtyard . . ."] (*Obras completas, 1923–1972* 627–28 [Hurley 285]).
6. "El libro de arena" ["The Book of Sand"]
   i. "Me dijo que su libro se llamaba el Libro de Arena, porque ni el libro ni la arena tienen ni principio ni fin" ["He told me his book was called the Book of Sand because neither sand nor this book has a beginning or an end"] (*Obras completas, 1975–1985* 69 [Hurley 481]).

- One object, symbol, or entity has many random guises but is in fact one and the same in essence
  1. "Tlön, Uqbar, Orbis Tertius"
     i. ". . . que tal dolor, que tal matiz verdoso del Amarillo, que tal temperature, que tal sonido, son la única realidad" [". . . a certain pain, a certain greenish-yellow color, a certain temperature, and a certain sound are all the same, single reality"] (*Obras completas, 1923–1972* 438n1 [Hurley 76n4]).
  2. "El Zahir" ["The Zahir"]
     i. "En Buenos Aires el Zahir es una moneda común. . . . (En Guzerat . . . un tigre fue Zahir; en Java, un ciego de la mezquita de Surakarta . . . en Persia, un astrolabio . . . una pequeña brújula . . . una veta en el mármol de uno de los mil doscientos pilares; en la judería de Tetuán, el fondo de un pozo.)" ["In Buenos Aires the Zahir is a common twenty-centavo coin. . . . (In Gujarat . . . the Zahir was a tiger; in Java it was a blind man in the Surakarta mosque . . . in Persia, an astrolabe . . . a small sailor's compass . . . a vein in the marble of one of the twelve hundred pillars; in the ghetto in Tetuán, the bottom of a well.)"] (*Obras completas, 1923–1972* 589 [Hurley 242]).

II. (De)Gradation

- (De)Gradation of ideal forms; categories; representations
  1. "Tlön, Uqbar, Orbis Tertius"
     - i. "los *hrönir* de segundo y de tercer grado . . . exageran las aberraciones del inicial. . . . el *hrön* de duodécimo grado ya empieza a decaer" ["*hrönir* of the second and third remove . . . exaggerate the aberrations of the first. . . . The *hrönir* of the twelfth remove begin to degenerate"] (*Obras completas, 1923–1972* 440 [Hurley 78]).
  2. "La biblioteca de Babel" ["The Library of Babel"]
     - i. ". . . cada ejemplar es único, irreemplazable, pero (como la Biblioteca es total) hay siempre varios centenares de miles de facsímiles imperfectors: de obras que no difieren sino por una letra o por una coma" [". . . each book is unique and irreplaceable, but (since the Library is total) there are always several hundred thousand imperfect facsimiles—books that differ by no more than a single letter, or a comma"] (*Obras completas, 1923–1972* 469 [Hurley 116]).
  3. "El Congreso" ["The Congress"]
     - i. "Planear una asamblea que representara a todos los hombres era como fijar el número exacto de los arquetipos platónicos. . . . don Alejandro Glencoe podía representar a los hacendados, pero también a los orientales y también a los grandes precursores y también a los hombres de barba roja y a los que están sentados en un sillón. Nora Erfjord era noruega. ¿Representaría a las secretarias, a las noruegas o simplemente a todas las mujeres hermosas? ¿Bastaba un ingeniero para representar a todos los Ingenieros, incluso los de Nueva Zelandia?" ["Designing a body of men and women which would represent all humanity was akin to fixing the exact number of Platonic archetypes. . . . don Alejandro Glencoe might represent ranchers, but also Uruguayans, as well as founding fathers and red-bearded men and men sitting in armchairs. Nora Erfjord was Norwegian. Would she represent secretaries, Norwegians, or simply all beautiful women? Was one engineer sufficient to represent all engineers, even engineers from

New Zealand?"] (*Obras completas, 1975–1985* 24 [Hurley 426–27]).

- A 1:1 map-territory relation
    1. "Pierre Menard, autor del Quijote" ["Pierre Menard, Author of the *Quixote*"]
        i. "Su admirable ambición era producer unas páginas que coincidieran—palabra por palabra—y línea por línea—con las de Miguel de Cervantes" ["His admirable position was to produce a number of pages which coincided—word for word and line for line—with those of Miguel de Cervantes"] (*Obras completas, 1923–1972* 446 [Hurley 91]).
    2. "Funes el memorioso" ["Funes, His Memory"]
        i. ". . . había reconstruido un día entero; no había dudado nunca, pero cada reconstrucción había requerido un día entero" [". . . he had constructed an entire day; he had never once erred or faltered, but each reconstruction had itself taken an entire day"] (*Obras completas, 1923–1972* 488 [Hurley 135]).
    3. "Tema del traidor y del héroe" ["The Theme of the Traitor and the Hero"]
        i. ". . . los *Festspiele* de Suiza; vastas y errantes representaciones teatrales, que requieren miles de actors y que reiteran episodios históricos en las mismas ciudades y montañas donde ocurrieron" [". . . the Swiss *Festspiele*—vast peripatetic theatrical performances that require thousands of actors and retell historical episodes in the same cities, the same mountains in which they occurred"] (*Obras completas, 1923–1972* 497 [Hurley 144]).
    4. "Del rigor en la ciencia" ["On Exactitude in Science"]
        i. ". . . los Colegios de Cartógrafos levantaron un Mapa del Imperio, que tenía el tamaño del Imperio y coincidía puntualmente con él" [". . . the Cartographers Guilds struck a Map of the Empire whose size was that of the Empire, and which coincided point for point with it"] (*Obras completas, 1923–1972* 847 [Hurley 325]).
    5. "El Congreso" ["The Congress"]
        i. In an afterword, Borges writes about "The Congress" that "su tema es una empresa tan vasta que se confunde al fin con el cosmos y con la suma de los días" ["its subject is a

company so vast that it merges at last into the cosmos itself and into the sum of days"] (*Obras completas, 1975–1985* 72 [Hurley 484]).

- A suggestion or explicit mention of infinite recursion or regression
  1. "El acercamiento a Almotásim" ["The Approach to Al-Mu'tasim"]
     i. ". . . la conjetura de que también el Todopoderoso está en busca de Alguien, y ese Alguien de Alguien superior (o simplemente imprescindible e igual) y así hasta el Fin—o major, el Sinfín—del Tiempo, o en forma cíclica" [". . . the idea that the Almighty is also in search of Someone, and that Someone, in search of a yet superior (or perhaps simply necessary, albeit equal) Someone, and so on, to the End—or better yet, the Endlessness—of Time. Or perhaps cyclically"] (*Obras completas, 1923–1972* 417 [Hurley 85]).
  2. "Las ruinas circulares" ["The Circular Ruins"]
     i. "No ser un hombre, ser la proyección del sueño de otro hombre ¡qué humillación incomparable, qué vertigo!" ["To be not a man, but the projection of another man's dream—what incomparable humiliation, what vertigo!"] (*Obras completas, 1923–1972* 454 [Hurley 100]).
  3. "La biblioteca de Babel" ["The Library of Babel"]
     i. ". . . el evangelio gnóstico de Basílides, el comentario de ese evangelio, el comentario del comentario de ese evangelio . . ." [". . . the gnostic gospel of Basilides, the commentary upon that gospel, the commentary on the commentary on that gospel . . ."] (*Obras completas, 1923–1972* 468 [Hurley 115]).
     ii. "Alguien propuso un método regresivo: Para localizer el libro A, consultar previamente un libro B que indique el sitio de A; para localizar el libro B, consultar previamente un libro C, y así hasta lo infinito . . ." ["Someone proposed searching by regression: To locate book A, first consult book B, which tells where book A can be found; to locate book B, first consult book C, and so on, to infinity . . ."] (*Obras completas, 1923–1972* 469 [Hurley 117]).
  4. "El jardín de senderos que se bifurcan" ["The Garden of Forking Paths"]

   i. "... esa noche que está en el centro de las 1001 Noches, cuando la reina Shahrazad (por una mágica distracción del copista) se pone a referir textualmente la historia de las 100 Noches, con riesgo de llegar otra vez a la noche en que la refiere, y así hasta lo infinito" ["... that night at the centre of the *1001 Nights,* when the queen Scheherazade (through some magical distractedness on the part of the copyist) begins to retell, verbatim, the story of the 1001 Nights, with the risk of returning once again to the night on which she is telling it—and so on, *ad infinitum*"] (*Obras completas, 1923–1972* 477 [Hurley 125]).

5. "La busca de Averroes" ["Averroës' Search"]
   i. "... mi narración era un símbolo del hombre que yo fui, mientras la escribía y que, para redactar esa narración, yo tuve que ser aquel hombre y que, para ser aquel hombre, yo tuve que redactar esa narración, y así hasta lo infinito" ["... my story was a symbol of the man I had been as I was writing, and that in order to write that story I had had to be that man, and that in order to be that man I had had to write that story, and son on, *ad infinitum*"] (*Obras completas, 1923–1972* 588 [Hurley 241]).
6. "El Aleph" ["The Aleph"]
   i. "... vi en el Aleph la tierra, y en la tierra otra vez el Aleph y en el Aleph la tierra ..." ["... saw the earth in the Aleph and the Aleph in the Aleph, and the Aleph once more in the earth and the earth in the Aleph ..."] (*Obras completas, 1923–1972* 626 [Hurley 283–84]).
7. "*Everything and Nothing*"
   i. "*... yo soñé el mundo como tú soñaste tu obra, mi Shakespeare, y entre las formas de mi sueño estás tú, que como yo eres muchos y nadie*" ["*... I dreamed the world as you, Shakespeare, dreamed your own work, and among the forms of my dream are you, who like me are many, yet no one*"] (*Obras completas, 1923–1972* 804 [Hurley 320]).
8. "Magias parciales del Quijote" ["Partial Magic in the *Quixote*"]
   i. "... la noche DCII, mágica entre las noches. En esa noche, el rey oye de boca de la reina su propia historia. Oye el principio de la historia, que abarca a todas las demás, y

también—de monstruoso modo—, a sí misma" [". . . the six hundred and second night, magical among all the nights. On that night, the king hears from the queen his own story. He hears the beginning of the story, which comprises all the others and also—monstrously—itself"] (*Obras completas, 1923–1972* 668–69 [Irby 195]).

- Doubt about an entity's—or one's own—ontological status; fiction overtaking reality; the imagined imposed onto reality
  1. "Tlön, Uqbar, Orbis Tertius"
     - i. "Tal fue la primera intrusión del mundo fantástico en elm undo real" ["That was the first intrusion of the fantastic world of Tlön into the real world"] (*Obras completas, 1923–1972* 441 [Hurley 80]).
     - ii. "ya en las memorias un pasado ficticio ocupa el sitio de otro, del que nada sabemos con certidumbre—ni siquiera que es falso" [". . . already a fictitious past has supplanted in men's memories that other past, of which we now know nothing is certain—not even that it is false"] (*Obras completas, 1923–1972* 443 [Hurley 81]).
     - iii. "El mundo será Tlön" ["The world will be Tlön"] (*Obras completas, 1923–1972* 443 [Hurley 81]).
  2. "Las ruinas circulares" ["The Circular Ruins"]
     - i. "quería soñarlo con integridad minuciosa e imponerlo a la realidad" ["He wanted to dream him completely, in painstaking detail, and impose him upon reality"] (*Obras completas, 1923–1972* 451 [Hurley 97]).
     - ii. ". . . comprendió que él también era una apariencia, que otro estaba soñándolo" [". . . he realized that he, too, was but appearance, that another man was dreaming him"] (*Obras completas, 1923–1972* 455 [Hurley 100]).
  3. "Tema del traidor y del héroe" ["The Theme of the Traitor and the Hero"]
     - i. "Kilpatrick fue ultimado en un teatro, pero de teatro hizo también la entera ciudad, y los actors fueron legión, y el drama coronado por su muerte abarcó muchos dias y muchas noches" ["Kilpatrick was murdered in a theater, yet the entire city played the role of theater, too, and the actors were legion, and the play that was crowned by Kilpatrick's death took place over many days and many nights"] (*Obras completas, 1923–1972* 497 [Hurley 145]).

4. "El Zahir" ["The Zahir"]
    i. "Cuando todos los hombres de la tierra piensen, día y noche, en el Zahir, ¿cuál será un sueño y cuál una realidad, la tierra o el Zahir?" ["When every man on earth thinks, day and night, of the Zahir, which will be dream and which reality, the earth or the Zahir?"] (*Obras completas, 1923–1972* 595 [Hurley 248–49]).

- Anachronistic precursors
    1. "Pierre Menard, autor del Quijote" ["Pierre Menard, Author of the *Quixote*"]
        i. ". . . nos insta a recorrer la Odisea como si fuera posterior a la Eneida . . ." [". . . encourages us to read the *Odyssey* as though it came after the *Æneid* . . ."] (*Obras completas, 1923–1972* 450 [Hurley 95]).
    2. "Kafka y sus precursores" ["Kafka and His Precursors"]
        i. "En cada uno de esos textos está la idiosincrasia de Kafka, en grado mayor o menor, pero si Kafka no hubiera escrito, no la percibiríamos; vale decir, no existiría. El poema *Fears and Scruples* de Robert Browning profetiza la obra de Kafka, pero nuestra lectura de Kafka afina y desvía sensiblemente nuestra lectura del poema. Browning no lo leía como ahora nosotros lo leemos. . . . El hecho es que cada escritor *crea* a sus precursores. Su labor modifica nuestra concepción del pasado, como ha de modificar el futuro" ["Kafka's idiosyncrasy is present in each of these writings, to a greater or lesser degree, but if Kafka had not written, we would not perceive it; that is to say, it would not exist. The poem 'Fears and Scruples' by Robert Browning prophesies the work of Kafka, but our reading of Kafka noticeably refines and diverts our reading of the poem. Browning did not read it as we read it now. . . . The fact is that each writer *creates* his precursors. His work modifies our conception of the past, as it will modify the future"] (*Obras completas, 1923–1972* 711, 712 [Weinberger 365].

III. Variance

- Permutation
    1. "Tlön, Uqbar, Orbis Tertius"

   i. "Los de ficción abarcan un solo argumento, con todas las permutaciones imaginables" ["Their fiction has but a single plot, with every imaginable permutation"] (*Obras completas, 1923–1972* 439 [Hurley 77]).
2. "La biblioteca de Babel" ["The Library of Babel"]
   i. ". . . nociones de análisis combinatiorio, ilustradas por ejemplos de variaciones con repetición ilimitada" [". . . the rudiments of combinatory analysis, illustrated with examples of endlessly repeating variations"] (*Obras completas, 1923–1972* 467 [Hurley 114]).
   ii. ". . . sus anaqueles registran todas las posibles combinaciones de los veintitantos simbolos ortográficos . . ." [". . . its bookshelves contain all possible combinations of the twenty-two orthographic symbols . . ."] (*Obras completas, 1923–1972* 467 [Hurley 115]).
3. "El jardín de senderos que se bifurcan" ["The Garden of Forking Paths"]
   i. "Naturalmente, hay varios desenlaces posibles. . . . En la obra de Ts'ui Pên, todos los desenlaces ocurren; cada uno es el punto de partida de otras bifurcaciones" ["Naturally, there are various possible outcomes. . . . In Ts'ui Pen's novel, *all* the outcomes in fact occur; each is the starting point for further bifurcations"] (*Obras completas, 1923–1972* 478 [Hurley 125]).
4. "Funes el memorioso" ["Funes, His Memory"]
   i. ". . . Funes no sólo recordaba cada hoja de cada árbol de cada monte, sino cada una de las veces que la había percibido o imaginado" [". . . Funes remembered not only every leaf of every tree in every patch of forest, but every time he had perceived or imagined that leaf"] (*Obras completas, 1923–1972* 489 [Hurley 136]).
5. "El inmortal" ["The Immortal"]
   i. ". . . pensé que nuestras percepciones eran iguales, pero que Argos las combinaba de otra manera y construia con ellas otros objetos . . ." [". . . I reflected that our perceptions were identical but that Argos combined them differently than I, constructed from them different objects . . ."] (*Obras completas, 1923–1972* 539 [Hurley 189]).
6. "Biografía de Tadeo Isidoro Cruz (1829–1874)" ["A Biography of Tadeo Isidoro Cruz (1829–1974)"]

   i. "La aventura consta en un libro insigne; es decir, en un libro cuya materia puede ser todo para todos (I Corintios 9:22), pues es capaz de casi inagotables repeticiones, versiones, perversiones" ["The adventure is recorded in a very famous book—that is, in a book whose subject can be all things to all men (I Corinthians 9:22), for it is capable of virtually inexhaustible repetitions, versions, perversions"] (*Obras completas, 1923–1972* 561 [Hurley 212]).

7. "La escritura del Dios" ["The Writing of the God"]
   i. "Dediqué largos años a aprender el orden y la configuración de las manchas. . . . las negras formas que tachaban el pelaje amarillo. Algunas incluían puntos; otras formaban rayas traversales en la cara interior de las piernas; otras, anulares, se repetían. Acaso eran un mismo sonido o una misma palabra" ["Long years I devoted to learning the order and arrangement of the spots on a tiger's skin. . . . the black shapes that mottled the yellow skin. Some made circles; others formed transverse stripes on the inside of its legs; others, ringlike, occurred over and over again—perhaps they were the same sound, or the same word"] (*Obras completas, 1923–1972* 597 [Hurley 251–52]).
8. "La trama" ["The Plot"]
   i. "Al destino le agradan las repeticiones, las variantes, las simetrías . . ." ["Fate is partial to repetitions, variations, symmetries"] (*Obras completas, 1923–1972* 793 [Hurley 307]).

- The same words are different depending on who expresses them
  1. "Pierre Menard, autor del Quijote" ["Pierre Menard, Author of the *Quixote*"]
     i. "Ser, de alguna manera, Cervantes y llegar al Quijote le pareció menos arduo—por consiguiente, menos interesante—que seguir siendo Pierre Menard y llegar al Quijote, a través de las experiencias de Pierre Menard" ["Being, somehow, Cervantes, and arriving thereby at the Quixote—that looked to Menard less challenging (and therefore less interesting) than continuing to be Pierre Menard and coming to the Quixote *through the experiences of Pierre Menard*"] (*Obras completas, 1923–1972* 447 [Hurley 91]).
     ii. ". . . el fragmentario Quijote de Menard es más sutil que el de Cervantes" [". . . Menard's fragmentary Quixote is more

subtle than Cervantes'"] (*Obras completas, 1923–1972* 448 [Hurley 93]).

iii. "El texto de Cervantes y el de Menard son verbalmente idénticos, pero el segundo es casi infinitamente más rico" ["The Cervantes text and the Menard text are verbally identical, but the second is almost infinitely richer"] (*Obras completas, 1923–1972* 449 [Hurley 94]).

2. "El inmortal" ["The Immortal"]

   i. "Los que siguen son más curiosos. Una oscura razón elemental me obligó a registrarlos; lo hice porque sabia que eran patéticos. No lo son, dichos por el romano Flaminio Rufo. Lo son, dichos por Homero . . ." ["The 'facts' that follow are even more curious. A dark yet elemental reason led me to put them to paper: I knew they were pathetic. They are not pathetic when narrated by the Roman Flaminius Rufus; they are when narrated by Homer"] (*Obras completas, 1923–1972* 543 [Hurley 194]).

- Entities are different at different moments

  1. "Funes el memorioso" ["Funes, His Memory"]

     i. "No sólo le costaba comprender que el símbolo genérico *perro* abarcara tantos individuos dispares de diversos tamaños y diversa forma; le molestaba que el perro de las tres y catorce (visto de perfil) tuviera el mismo nombre que el perro de las tres y cuarto (visto de frente)" ["Not only was it difficult for him to see that the generic symbol 'dog' took in all the dissimilar individuals of all shapes and sizes, it irritated him that the 'dog' of three-fourteen in the afternoon, seen in profile, should be indicated by the same noun as the dog of three-fifteen, seen frontally"] (*Obras completas, 1923–1972* 490 [Hurley 136]).

  2. "La busca de Averroes" ["Averroës' Search"]

     i. ". . . la luna de Bengala no es igual a la luna del Yemen, pero se deja describir con las mismas voces" [". . . the moon of Bengal is not the same as the moon of Yemen, but it deigns to be described with the same words"] (*Obras completas, 1923–1972* 584–85 [Hurley 238]).

IV. Systematization

- Classification and taxonomy; the invention of a system for such

1. "La biblioteca de Babel" ["The Library of Babel"]
    i. ". . . el catálogo fiel de la Biblioteca, miles y miles de catálogos falsos, la demostración de la falacia de esos catálogos, la demostración de la falacia del catálogo verdadero . . ." [". . . the faithful catalog of the Library, thousands and thousands of false catalogs, the proof of the falsity of those false catalogs, the proof of the falsity of the *true* catalog . . ."] (*Obras completas, 1923–1972* 467–68 [Hurley 115]).
2. "Funes el memorioso" ["Funes, His Memory"]
    i. ". . . había discurrido un sistema original de numeración. . . . Cada palabra tenía un signo particular, una especie de marca . . . un vocabulario infinito para la serie natural de los números . . ." [". . . he had invented a numbering system original with himself. . . . Every word had a particular figure attached to it, a sort of marker . . . an infinite vocabulary for the natural series of numbers . . ."] (*Obras completas, 1923–1972* 489 [Hurley 135, 136]).
    ii. "Resolvió reducir cada una de sus jornadas pretéritas a unos setenta mil recuerdos, que definiría luego por cifras. . . . Pensó que en la hora de la muerte no habría acabado aún de clasificar todos los recuerdos de la niñez. . . . un inútil catálogo mental de todas las imágenes del recuerdo . . ." ["He resolved to reduce every one of his past days to some seventy thousand recollections, which he would then define by numbers. . . . He saw that by the time he died he would still not have finished classifying all the memories of his childhood. . . . a pointless mental catalog of all the images of his memory . . ."] (*Obras completas, 1923–1972* 489 [Hurley 136]).
3. "El Congreso" ["The Congress"]
    i. ". . . don Alejandro concibió el propósito de organizer un Congreso del Mundo que representaría a todos los hombres de todas las naciones" [". . . don Alejandro conceived the idea of establishing a Congress of the World, which would represent all people of all nations"] (*Obras completas, 1975–1985* 23 [Hurley 426]).
4. "El idíoma analítico de John Wilkins" ["John Wilkins' Analytical Language"]

i. "Dividió el universo en cuarenta categorías o géneros, subdivisibles luego en diferencias, subdivisibles a su vez en especies" ["He divided the universe into forty categories or classes, which were then subdivided into differences, and subdivided in turn into species"] (*Obras completas, 1923–1972* 707 [Weinberger 230]).

ii. ". . . notoriamente no hay clasificación del universo que no sea arbitraria y conjectural" [". . . obviously there is no classification of the universe that is not arbitrary and speculative"] (*Obras completas, 1923–1972* 708 [Weinberger 231]).

- Strange language; the creation or existence of a (universal) language
  1. "Tlön, Uqbar, Orbis Tertius"
     i. "No hay sustantivos en la conjectural *Ursprache* de Tlön . . ." ["There are no nouns in the conjectural *Ursprache* of Tlön . . ."] (*Obras completas, 1923–1972* 435 [Hurley 73]).
     ii. "En los del hemisferio boreal . . . la célula primordial no es el verbo, sino el adjetivo monosilábico. El sustantivo se forma por acumulación de adjetivos" ["In the northern hemisphere . . . the primary unit is not the verb but the monosyllabic adjective. Nouns are formed by stringing together adjectives"] (*Obras completas, 1923–1972* 435 [Hurley 73]).
  2. "Pierre Menard, autor del Quijote" ["Pierre Menard, Author of the *Quixote*"]
     i. "Un monografía sobre la posibilidad de construir un vocabulario poético de conceptos que no fueran sinónimos o perífrasis de los que informan el lenguaje común . . ." ["a monograph on the possibility of constructing a poetic vocabulary from concepts that are neither synonyms nor periphrastic locutions for the concepts that inform common speech . . ."] (*Obras completas, 1923–1972* 444 [Hurley 89]).
  3. "Funes el memorioso" ["Funes, His Memory"]
     i. ". . . un idioma imposible en el que cada cosa individual, cada piedra, cada pájaro y cada rama tuviera un nombre propio . . ." [". . . an impossible language in which each individual thing—every stone, every bird, every branch—would have its own name . . ."] (*Obras completas, 1923–1972* 489 [Hurley 136]).

4. "El inmortal" ["The Immortal"]
   i. ". . . consideré la posibilidad de un lenguaje que ignoraba los sustantivos, un lenguaje de verbos impersonables o de indeclinables epítetos" [". . . I toyed with the possibility of a language that had no nouns, a language of impersonal verbs or indeclinable adjectives"] (*Obras completas, 1923–1972* 539 [Hurley 189]).
5. "El Congreso" ["The Congress"]
   i. "No descuidé las lenguas universales; me asomé al esperanto . . . y al Volapük, que quiere explorer todas las posibilidades lingüísticas, declinando los verbos y conjugando los sustantivos. . . . Me demoré asimismo en el examen del idioma analítico de John Wilkins, donde la definición de cada palabra está en las letras que la forman" ["I did not neglect the universal languages; I looked into Esperanto . . . and Volapük, which attempts to explore all the possibilities of language, declining verbs and conjugating nouns. . . . I also devoted some time to the study of the analytical language of John Wilkins, in which the definition of each word is contained in the letters that constitute it"] (*Obras completas, 1975–1985* 28 [Hurley 431–32]).
6. "El idíoma analítico de John Wilkins" ["John Wilkins' Analytical Language"]
   i. "En el idioma universal que ideó Wilkins al promediar el siglo XVII, cada palabra se define a sí misma" ["In the universal language conceived by Wilkins in the middle of the seventeenth century, each word defines itself"] (*Obras completas, 1923–1972* 706 [Weinberger 230]).

- Inventorying an author's works
  1. "Pierre Menard, autor del Quijote" ["Pierre Menard, Author of the *Quixote*"]
     i. "Examinado con esmero su archivo particular, he verificado que consta de las piezas que siguen . . ." ["Having examined his personal files with the greatest care, I have established that his body of work consists of the following pieces . . ."] (*Obras completas, 1923–1972* 444 [Hurley 88]).
  2. "Examen de la obra de Herbert Quain" ["A Survey of the Works of Herbert Quain"]
     i. The story's title

WORKS CITED

Abate, Michelle Ann. "Taking Silliness Seriously: Jim Henson's *The Muppet Show*, the Anglo-American Tradition of Nonsense, and Cultural Critique." *The Journal of Popular Culture*, vol. 42, no. 4, 2009, pp. 589–613.

Abbott, Edwin A. *Flatland: A Romance of Many Dimensions*. Dover Publications, 1992.

Abbott, H. Porter. *The Cambridge Introduction to Narrative*. Cambridge UP, 2021.

Abbott, H. Porter. "Narrative and Emergent Behavior." *Poetics Today*, vol. 29, no. 2, 2008, pp. 227–44.

"About Us." *Friends of Pando*. https://www.friendsofpando.org/about-us/. Accessed 1 Jan. 2026.

Abrams, M. H. *The Mirror and the Lamp: Romantic Theory and the Critical Tradition*. Oxford UP, 1953.

Aizenberg, Edna. "Introduction." *Borges and His Successors: The Borgesian Impact on Literature and the Arts*, U of Missouri P, 1990, pp. 1–14.

Alazraki, Jaime. "*62, modelo para armar*: Novela calidoscopio." *Revista Iberoamericana*, vol. 47, nos. 116 17, 1981, pp. 155–63.

Alazraki, Jaime. "Borges' Modernism and the New Critical Idiom." *Borges and the Kabbalah*, Cambridge UP, 1988, pp. 166–75.

Alazraki, Jaime. "Toward the Last Square of the Hopscotch." *Julio Cortázar*, edited by Harold Bloom, Chelsea House Publishers, 2005, pp. 3–26.

Alber, Jan. "Unnatural Narratology: The Systematic Study of Anti-Mimeticism." *Literature Compass*, vol. 10, no. 5, 2013, pp. 449–60.

Alber, Jan, et al. "What Really Is Unnatural Narratology?" *Storyworlds: A Journal of Narrative Studies*, vol. 5, 2013, pp. 101–18. https://doi.org/10.5250/storyworlds.5.2013.0101.

Aldama, Frederick Luis, and James Phelan. "Episode 11." *Project Narrative Podcast*, The Ohio State University, 27 Aug. 2022. https://podcast.osu.edu/projectnarrative/2022/08/27/episode-11-jim-phelan-frederick-aldama-the-continuity-of-parks-by-julio-cortazar/.

"Aleatory." *The Oxford Dictionary of Literary Terms*. Chris Baldick, Oxford UP, 2015. https://www.oxfordreference.com/view/10.1093/acref/9780198715443.001.0001/acref-9780198715443-e-26. Accessed 30 Dec. 2025.

"Andy Samberg / The Muppets / Final Show." *Late Night with Jimmy Fallon*, NBC, 2014.

"Anomalous, *Adj.*" *Collins English Dictionary*, HarperCollins Publishers. https://www.collinsdictionary.com/dictionary/english/anomalous. Accessed 27 Feb. 2025.

"Anomalous, *Adj.*" *Kernerman Webster's College Dictionary*, Random House, 2010. https://www.thefreedictionary.com/anomalous. Accessed 27 Feb. 2025.

"Anthology, *N.*" *Online Etymology Dictionary*. https://www.etymonline.com/word/anthology. Accessed 12 Mar. 2026.

Asimov, Isaac. *I, Robot*. Del Rey, 2008.

Atwood, Margaret. *The Blind Assassin*. Nan A. Talese, 2000.

Atwood, Margaret. *Oryx and Crake: A Novel*. Nan A. Talese, 2003.

Baake, Ken. *Metaphor and Knowledge: The Challenges of Writing Science*. SUNY P, 2003.

Bal, Mieke. *Narratology: Introduction to the Theory of Narrative*. Translated by Christine van Boheemen, U of Toronto P, 2009.

Balderston, Daniel. *The Literary Universe of Jorge Luis Borges*. Greenwood Press, 1986.

Banchoff, Thomas F. "The Axioms of Euclidean Plane Geometry." *Beyond the Third Dimension*. https://www.math.brown.edu/tbanchof/Beyond3d.new/chapter9/s9_1.html. Accessed 27 Feb. 2025.

Barry, Andrew, et al. "Logics of Interdisciplinarity." *Economy and Society*, vol. 37, no. 1, 2008, pp. 20–49.

Bartlett, Catherine. "Magical Realism: The Latin American Influence on Modern Chicano Writers." *Confluencia*, vol. 1, no. 2, 1986, pp. 27–37.

Barthes, Roland. "Theory of the Text." *Untying the Text: A Post-Structuralist Reader*, edited by Robert Young, Routledge, 1981, pp. 31–47.

Bejan, Adrian, and J. Peder Zane. *Design in Nature: How the Constructal Law Governs Evolution in Biology, Physics, Technology, and Social Organization*. Doubleday, 2012.

Benyus, Janine M. *Biomimicry: Innovation Inspired by Nature*. William Morrow, 1997.

Berveiller, Emile. *Le Cosmopolitisme de Jorge Luis Borges*. Publications de la Sorbonne, 1973.

Black, Max. *Models and Metaphors: Studies in Language and Philosophy*. Cornell UP, 1962.

Bloom, Harold. *Genius: A Mosaic of One Hundred Exemplary Creative Minds*. Warner Books, 2002.

Bloom, Harold. *How to Read and Why*. Scribner, 2000.

Boldy, Steven. *The Novels of Julio Cortázar*. Cambridge UP, 1980.

Borgenicht, David. *Sesame Street Unpaved*. Hyperion Publishing, 1998.

Borges, Jorge Luis. *Collected Fictions*. Edited and translated by Andrew Hurley, Penguin Books, 1999.

Borges, Jorge Luis. "Elementos de preceptiva." *Sur*, no. 7, 1933, pp. 158–61.

Borges, Jorge Luis. *Labyrinths: Selected Stories and Other Writings*. Edited by Donald A. Yates and James E. Irby, New Directions, 2007.

Borges, Jorge Luis. *Obras completas, 1923–1972*. Emecé Editores, 1989.

Borges, Jorge Luis. *Obras completas, 1975–1985*. Emecé Editores, 1989.

Borges, Jorge Luis. *Selected Non-Fictions.* Edited by Eliot Weinberger, translated by Esther Allen, Suzanne Jill Levine, and Eliot Weinberger, Penguin Books, 1999.

Brink, André. *The Novel: Language and Narrative from Cervantes to Calvino.* New York UP, 1998.

Bronowski, J. *Science and Human Values.* Harper & Row, 1965.

Burgass, Catherine. "A Brief Story of Postmodern Plot." *The Yearbook of English Studies,* vol. 30, 2000, pp. 177–86.

Burke, Kenneth. *Permanence and Change: An Anatomy of Purpose.* Bobbs-Merrill, 1965.

Butler, Octavia E. *Dawn: A Lilith's Brood Novel.* Grand Central Publishing, 2021.

Cabiya, Pedro. *Malas hierbas.* Zemi Book, 2011.

Cabiya, Pedro. *Wicked Weeds: A Novel.* Translated by Jessica Ernst Powell, Mandel Vilar Press, 2016.

Cajetan, Tommaso de Vio. *"The Analogy of Names" and "The Concept of Being."* Translated by Edward A. Bushinski and Henry J. Koren, Duquesne U, 1953.

Campbell, Joseph. *The Hero with a Thousand Faces.* New World Library, 2008.

Caracciolo, Marco. *Narrating the Mesh: Form and Story in the Anthropocene.* U of Virginia P, 2021.

Carey, Stan. "Linguistic Sci-Fi in *Embassytown.*" *Sentence First,* 29 Aug. 2017. https://stancarey.wordpress.com/2017/08/29/linguistic-sci-fi-in-embassytown/.

Carlisle, Alli. "The Constructed Island: Experimentation and Reference in Ana Menéndez's *Adios, Happy Homeland!*" *Critical Insights: Contemporary Immigrant Short Fiction,* edited by Robert C. Evans, Salem Press, 2015, pp. 165–80.

Carpentier, Alejo. "Viaje a la semilla." *Guerra del tiempo y otros relatos.* Lectorum, 2003, pp. 55–72.

Carrillo, Natalia, and Sergio Martínez. "Scientific Inquiry: From Metaphors to Abstraction." *Perspectives on Science,* vol. 31, no. 2, 2023, pp. 233–61.

Carroll, Lewis. *Alice's Adventures in Wonderland.* Macmillan & Co., 1865.

"Causality." *The Oxford Companion to Cosmology,* edited by Andrew Liddle and Jon Loveday, Oxford UP, 2008. https://www.oxfordreference.com/view/10.1093/acref/9780198608585.001.0001/acref-9780198608585-e-42. Accessed 29 Aug. 2025.

"Chapter 15." *House of Cards,* written by Beau Willimon, directed by Carl Franklin, Netflix, 2014.

Charap, John M. *Explaining the Universe: The New Age of Physics.* Princeton UP, 2004.

Chiang, Ted. *Stories of Your Life and Others.* Vintage Books, 2016.

Christ, Ronald. *The Narrow Act: Borges' Art of Allusion.* Lumen Books, 1995.

Cohn, Dorrit. "Metalepsis and Mise en Abyme." Translated by Lewis S. Gleich. *Narrative,* vol. 20, no. 1, 2012, pp. 105–14.

Cohn, Dorrit. "Signposts of Fictionality: A Narratological Perspective." *Poetics Today,* vol. 11, no. 4, 1979, pp. 775–804.

"Conan Interviews Kermit the Frog | Late Night with Conan O'Brien." *YouTube,* uploaded by Conan O'Brien, 18 June 2020. https://www.youtube.com/watch?v=YoVbUb_JXt4.

Coolidge, Frederick, et al. "Recursion: What Is It, Who Has It, and How Did It Evolve?" *WIREs Cognitive Science,* vol. 2, no. 5, 2011, pp. 547–54. https://doi.org/10.1002/wcs.131.

Cooper, Joel. *Cognitive Dissonance: Fifty Years of Classic Theory.* SAGE, 2007.

Copland, Sarah, and James Phelan. "The Ideal Narratee and the Rhetorical Model of Audiences." *Poetics Today,* vol. 43, no. 1, 2022, pp. 1–26.

Cortázar, Julio. *62/Modelo para armar.* Alfaguara, 1995.

Cortázar, Julio. *62: A Model Kit.* Translated by Gregory Rabassa, Marion Boyars Publishers, 1976.

Cortázar, Julio. "Continuidad de los parques." *Final del Juego,* Editorial Sudamericana, 1974, pp. 9–11.

Cortázar, Julio. "Continuity of Parks." *Blow-Up, and Other Stories,* translated by Paul Blackburn, Collier Books, 1968, pp. 55–56.

Cortázar, Julio. *Hopscotch.* Translated by Gregory Rabassa, Pantheon Books, 1966.

Cortázar, Julio. *Rayuela.* Edición conmemorativa, Real Academia Española: Asociación de Academias de la Lengua Española, 2019.

Covington, C. D. "The Signifier and the Signified: Semiotics and China Miéville's *Embassytown.*" *Reactor,* 9 Oct. 2019. https://reactormag.com/the-signifier-and-the-signified-semiotics-and-china-mievilles-embassytown/.

Dahl, Roald. *Charlie and the Chocolate Factory.* Knopf, 1964.

Dällenbach, Lucien. *The Mirror in the Text.* Translated by Jeremy Whiteley with Emma Hughes, U of Chicago P, 1989.

Davies, Paul. *The Goldilocks Enigma: Why Is the Universe Just Right for Life?* Mariner Books, 2008.

Davis, Michael. *Street Gang: The Complete History of Sesame Street.* Viking, 2008.

Daynes, Alistair. "The Complete History of Biomimicry (with Examples)." *Learn Biomimicry,* 1 Apr. 2025. https://www.learnbiomimicry.com/blog/history-of-biomimicry.

Defoe, Daniel. *Robinson Crusoe.* William Taylor, 1719.

Deleuze, Gilles, and Félix Guattari. *A Thousand Plateaus: Capitalism and Schizophrenia.* Translated by Brian Massumi, U of Minnesota P, 1987.

Delgado, Guillermo, and Bruno Estañol. "Acido, ergo sum: Holger Hydén—the Neuroscientist in Cortázar's Hopscotch." *Arquivos de neuro-psiquiatria,* vol. 71, no. 6, 2013, pp. 408–10. https://doi.org/10.1590/0004-282X20130048.

Dirac, Paul A. M. *The Principles of Quantum Mechanics.* Clarendon Press, 1958.

Dobrzynski, Dion. "Thinking Forests in Ursula K. Le Guin's 'Vaster Than Empires and More Slow' and *The Word for World Is Forest.*" *Journal of Literature and Science,* vol. 15, no. 1, 2022, pp. 44–58.

Doctor Who. Created by Sydney Newman, C. E. Webber, and Donald Wilson, BBC, 1963–present.

Duchesne-Sotomayor, Dafne. *¿Qué ocurre cuando el zombi habla? Lo posthumano en Malas hierbas de Pedro Cabiya.* Editora Educacion Emergente, 2024.

Echevarría, Roberto González. "*One Hundred Years of Solitude*: The Novel as Myth and Archive." *García Márquez,* edited by Robin W. Fiddian, Longman, 1995, pp. 79–99.

Eco, Umberto. *The Open Work.* Translated by Anna Cancogni, Hutchinson Radius, 1989.

Elen, Jan, et al. "Chapter 1: Introduction." *Links Between Beliefs and Cognitive Flexibility: Lessons Learned,* Springer, 2011.

"Ellen Meets Kermit!" *YouTube,* uploaded by TheEllenShow, 9 Nov. 2011. https://www.youtube.com/watch?v=1_AUn5poOmA.

"Elmo." *Instagram,* 6 Aug. 2022. https://www.instagram.com/elmo.

Ende, Michael. *The Neverending Story.* Translated by Ralph Manheim, Dutton Books, 1997.

"Epigenetics." *A Dictionary of Biomedicine.* Edited by John Lackie and Brian Nation, Oxford UP, 2019. https://www.oxfordreference.com/view/10.1093/acref/9780191829116.001.0001/acref-9780191829116-e-3214. Accessed 2 Jan. 2026.

"Explore Pando." *Friends of Pando.* https://www.friendsofpando.org/about-us/. Accessed 1 Jan. 2026.

"Fantastic, *Adj.*" *American Heritage Dictionary of the English Language,* HarperCollins Publishers, 2022. https://ahdictionary.com/word/search.html?q=fantastic. Accessed 28 Feb. 2025.

"Fantastic, *Adj.*" *Online Etymology Dictionary.* https://www.etymonline.com/search?q=fantastic. Accessed 28 Feb. 2025.

"Fantastic, *Adj.*" *Oxford Dictionary of English.* Edited by Angus Stevenson, Oxford UP, 2015. https://www.oxfordreference.com/view/10.1093/acref/9780199571123.001.0001/m_en_gb0287370. Accessed 30 Dec. 2025.

Farina, Lara. "Curl." *Veer Ecology: A Companion for Environmental Thinking,* edited by Jeffrey Jerome Cohen and Lowell Duckert, U of Minnesota P, 2017, pp. 434–54.

Federman, Raymond. *Take It or Leave It.* U of Alabama P, 1997.

Festinger, Leon. *A Theory of Cognitive Dissonance.* Stanford UP, 1962.

Feynman, Richard P., et al. "Quantum Behavior." *The Feynman Lectures on Physics,* vol. 3, California Institute of Technology, 2013, ch. 1. https://www.feynmanlectures.caltech.edu/III_01.html.

Ficek, Zbigniew, and Stuart Swain. *Quantum Interference and Coherence: Theory and Experiments.* Springer, 2005.

Fishburn, Evelyn. "A Footnote to Borges Studies: A Study of the Footnotes." *Occasional Papers,* U of London, 2002, pp. 1–23.

Fishburn, Evelyn, and Psiche Hughes. *A Dictionary of Borges.* Duckbacks, 1989.

Fludernik, Monika. *Towards a "Natural" Narratology.* Routledge, 1996.

Galindo Barrientos, Mayevi Hadith, and Lesbia María Orellana Duarte. "Epigenética: La magnitude del neurodesarrollo y la neuroembriología." *Revista Académica Sociedad del Conocimiento Cunzac,* vol. 2, no. 2, 2022, pp. 297–304.

Gammelgaard, Lasse R., et al. *Fictionality in Literature: Core Concepts Revisited.* The Ohio State UP, 2022.

García Márquez, Gabriel. *Cien años de soledad.* Edición conmemorativa, Real Academia Española: Asociación de Academias de la Lengua Española, 2007.

García Márquez, Gabriel. *One Hundred Years of Solitude.* Translated by Gregory Rabassa, Harper Perennial Modern Classics, 2006.

Geary, James. *I Is an Other: The Secret Life of Metaphor and How It Shapes the Way We See the World.* Harper Perennial, 2012.

Genette, Gérard. *Narrative Discourse: An Essay in Method.* Translated by Jane E. Lewin, Cornell UP, 1980.

Genette, Gérard. *The Architext: An Introduction.* Translated by Jane E. Lewin, U of California P, 1992.

Genette, Gérard. *Palimpsests: Literature in the Second Degree.* Translated by Channa Newman and Claude Doubinsky, U of Nebraska P, 1997.

Gentner, Dedre, and José Medina. "Similarity and the Development of Rules." *Cognition,* vol. 65, 1998, pp. 263–97.

Gleick, James. *The Information: A History, a Theory, a Flood.* Vintage, 2012.

Goble, Mark. "Convergence." *The Cambridge Companion to Twenty-First-Century American Fiction,* edited by Joshua L. Miller, Cambridge UP, 2021, pp. 215–33.

Godfrey-Smith, Peter. "Darwinian Individuals." *From Groups to Individuals: Evolution and Emerging Individuality,* edited by Frédéric Bouchard and Philippe Huneman, MIT P, 2013, pp. 17–36.

Gombrich, Carl. "An Outline of Alan Wildon's Philosophy of Knowledge, with Special Attention Given to the Importance of Interdisciplinarity and Its Role in Undergraduate Education." *Interdisciplinary Science Reviews*, vol. 44, nos. 3–4, 2019, pp. 342–60.

González, Aníbal. "Translation and Genealogy: *One Hundred Years of Solitude*." *Gabriel García Márquez: New Readings*, edited by Bernard McGuirk and Richard A. Cardwell, Cambridge UP, 1987, pp. 65–79.

"Google Street View Goes Inside Doctor Who Tardis." *BBC*, 14 Aug. 2013. https://www.bbc.co.uk/newsround/23698637.

Gorman, David. "Fiction, Theories Of." *Routledge Encyclopedia of Narrative Theory*, edited by David Herman et al., Routledge, 2008, pp. 163–67.

Grant, Michael C. "The Trembling Giant." *Discover Magazine*, 1 Oct. 1993. https://www.discovermagazine.com/the-trembling-giant-40013.

The Great Muppet Caper. Directed by Jim Henson. ITC Entertainment, Henson Associates, 1981.

Gribbin, John. *In Search of Schrödinger's Cat: Quantum Physics and Reality*. Bantam Books, 1984.

Hawking, Stephen, and Leonard Mlodinow. *The Grand Design*. Bantam Books, 2010.

Heisenberg, Werner. *The Physical Principles of the Quantum Theory*. Translated by Carl Eckart and Frank C. Hoyt, Dover Publications, 1949.

Hemingway, Ernest. *The Old Man and the Sea*. Charles Scribner's Sons, 1952.

Herman, David. *Storytelling and the Sciences of Mind*. MIT P, 2013.

Herman, David, et al., editors. *Routledge Encyclopedia of Narrative Theory*. Routledge, 2008.

Hoffman, Yonina. "*Infinite Jest*'s Reconfiguration: Delayed Disclosure and Systems Science." *Fictionality and Science in Dialogue*, special issue of *Style*, edited by Rhona Trauvitch and James Phelan, vol. 59, no. 3, 2025, pp. 437–65.

Hofstadter, Douglas R. *Gödel, Escher, Bach: An Eternal Golden Braid*. Harvester Press, 1979.

Hofstadter, Douglas R., and Emmanuel Sander. *Surfaces and Essences: Analogy as the Fuel and Fire of Thinking*. Basic Books, 2013.

Hoskins, Rachel. "Do Worms Have Eyes? And Other Worm Facts." *Woodland Trust*, 25 Apr. 2019. https://www.woodlandtrust.org.uk/blog/2019/04/do-worms-have-eyes/.

Hud, Nicholas V., et al. "The Origin of RNA and 'My Grandfather's Axe.'" *Chemistry & Biology*, vol. 20, 2013, pp. 466–74.

Husain, Tasneem Zehra. "Bridges to the Unknown." *New Humanist*, vol. 139, no. 1, 2024, pp. 22–25.

Interstellar. Directed by Christopher Nolan, Paramount Pictures, 2014.

Iser, Wolfgang. *The Act of Reading: A Theory of Aesthetic Response*. Johns Hopkins UP, 1978.

James, Henry. *The Turn of the Screw*. The Macmillan Company, 1898.

Jefferson, Ann. "'Mise en Abyme' and the Prophetic in Narrative." *Style*, vol. 17, no. 2, 1983, pp. 196–208.

"Johnny Carson." *Muppet Wiki*. https://muppet.fandom.com/wiki/Johnny_Carson. Accessed 1 Jan. 2026.

Jordan, Paul. "Rayuela: Novel by Julio Cortázar." *Encyclopedia of Latin American Literature*, edited by Verity Smith, Fitzroy Dearborn, 1997, pp. 227–29.

Joyce, James. *Finnegans Wake*. Faber and Faber, 1939.

"Keira Knightley / Tony Goldwyn / The Cookie Monster / Incubus." *Jimmy Kimmel Live!*, 12:05 Productions, 2015.

Kermit's Swamp Years. Directed by David Gumpel, Jim Henson Home Entertainment, 2002.

Kermode, Frank. *The Sense of an Ending: Studies in the Theory of Fiction: With a New Epilogue.* Oxford UP, 2000.

Kerr, Lucille. "Betwixt Reading and Repetition (Apropos of Cortázar's *62: A Model Kit*)." *Julio Cortázar,* edited by Harold Bloom, Chelsea House Publishers, 2005, pp. 227–44.

Kissel, Stephen C. "Teaching Matter and Phase Change." *Yale-New Haven Teachers Institute,* 2016. https://teachersinstitute.yale.edu/curriculum/units/2016/4/16.04.06/3. Accessed 26 May 2026.

Kövecses, Zoltán. *Metaphor: A Practical Introduction.* Oxford UP, 2010.

Kuhn, Thomas S. *The Structure of Scientific Revolutions.* U of Chicago P, 1996.

Kukkonen, Karin. "Metalepsis in Popular Culture: An Introduction." *Metalepsis in Popular Culture,* edited by Karin Kukkonen and Sonja Klimek, De Gruyter, 2011, pp. 1–21.

Lakoff, George, and Mark Johnson. *Metaphors We Live By.* U of Chicago P, 2003.

L'Engle, Madeline. *A Wrinkle in Time.* Ariel Books, 1962.

Le Guin, Ursula K. "Darkness Box." *The Wind's Twelve Quarters.* Harper & Row, 1975, pp. 60–70.

Le Guin, Ursula K. "Vaster Than Empires and More Slow." *The Wind's Twelve Quarters.* Harper & Row, 1975, pp. 181–217.

Lee, Harper. *To Kill a Mockingbird.* Grand Central Publishing, 1988.

Letter, Amy. "Escaping Isn't What It Used to Be: Review of Ana Menéndez's *Adios, Happy Homeland!*" *Anthurium: A Caribbean Studies Journal,* vol. 10, no. 1, 2013, pp. 1–6.

Lévi-Strauss, Claude. *Myth and Meaning: Cracking the Code of Culture.* Schocken, 1995.

Levine, Caroline. *Forms: Whole, Rhythm, Hierarchy, Network.* Princeton UP, 2015.

Lieber, Jeffrey, et al., creators. *Lost.* Grass Skirt Productions, Bad Robot, ABC Studios, 2010.

"Light Cone." *The Oxford Companion to Cosmology.* Edited by Andrew Liddle and Jon Loveday, Oxford UP, 2008. https://www.oxfordreference.com/view/10.1093/acref/9780198608585.001.0001/acref-9780198608585-e-226. Accessed 31 Aug. 2025.

Lucretius Carus, Titus. *On the Nature of Things.* Translated by Cyril Bailey, Clarendon Press, 1921.

Maguire, Emily A. "The Heart of a Zombie: Dominican Literature's Sentient Undead." *Alambique,* vol. 6, no. 1, 2018, pp. 1–20.

Mallarmé, Stéphane. *The Book.* Translated by Sylvia Gorelick, Exact Change, 2018.

Margolin, Uri. "Mathematics and Narrative: A Narratological Perspective." *Circles Disturbed: The Interplay of Mathematics and Narrative,* edited by Apostolos Doxiadis and Barry Mazur, Princeton UP, 2012, pp. 481–507.

McFadden, Johnjoe. *Quantum Evolution.* W. W. Norton, 2002.

McHale, Brian. *Postmodernist Fiction.* Routledge, 1987.

McManus, Elizabeth Berkebile. "Protecting the Island: Narrative Continuance in *Lost.*" *Journal of the Fantastic in the Arts,* vol. 22, no. 1, 2011, pp. 4–23.

"Medicine: The Chemistry of Thought." *Time,* 10 Feb. 1961. https://time.com/archive/6809768/medicine-the-chemistry-of-thought/.

Menéndez, Ana. *Adios, Happy Homeland!* Black Cat, 2011.

Merlin, Francesca. "Evolutionary Chance Mutation: A Defense of the Modern Synthesis' Consensus View." *Philosophy, Theory, and Practice in Biology,* vol. 2, no. 1, 2010. https://doi.org/10.3998/ptb.6959004.0002.003.

Meuter, Norbert. "Narration in Various Disciplines." *The Living Handbook of Narratology,* edited by Peter Hühn et al., Hamburg U, 2013.

Miéville, China. *Embassytown.* Del Ray Books, 2011.

"Miss Piggy Tells All About Kermit Break-Up." *Good Morning America*, 16 Sept. 2015. https://www.goodmorningamerica.com/news/video/miss-piggy-tells-kermit-break-33795043.

Mooij, J. J. A. *Fictional Realities: The Uses of Literary Imagination.* John Benjamins, 1993.

Moraru, Christian. "Intertextuality." *Routledge Encyclopedia of Narrative Theory*, edited by David Herman et al., Routledge, 2008, pp. 256–61.

Mr. Magorium's Wonder Emporium. Directed by Zach Helm. Mandate Pictures, Walden Media, FilmColony, 2007.

Mukherjee, Siddhartha. *The Gene: An Intimate History.* Scribner, 2016.

The Muppet Christmas Carol. Directed by Brian Henson. Walt Disney Pictures, Jim Henson Productions, 1992.

Muppets from Space. Directed by Tim Hill. Columbia Pictures, Jim Henson Pictures, 1999.

The Muppets Take Manhattan. Directed by Frank Oz. Henson Associates, 1984.

"My Whole Life Is Thunder." *30 Rock*, written by Jack Burditt and Colleen McGuinness, directed by Linda Mendoza, NBC, 2012.

Nabokov, Vladimir. *Pale Fire.* Vintage, 1989.

Nambisan, Padma. *An Introduction to Ethical, Safety and Intellectual Property Rights Issues in Biotechnology.* Academic Press, 2017.

National Academy of Sciences, National Academy of Engineering, and Institute of Medicine. *Facilitating Interdisciplinary Research.* National Academies Press, 2005. https://doi.org/10.17226/11153.

"Neil deGrasse Tyson." Superman Wiki. *Fandom.* https://superman.fandom.com/wiki/Neil_Degrasse_Tyson. Accessed 28 Feb. 2025.

Newman, Daniel Aureliano. "From Fi-Sci Pattern Mapping to Literary Interpretation: The Whodunit Plot and Quantum Superposition in David Lodge's *Thinks*. . . ." *Fictionality and Science in Dialogue*, special issue of *Style*, edited by Rhona Trauvitch and James Phelan, vol. 59, no. 3, 2025, pp. 311–35.

Nielsen, Henrik Skov, et al. "Ten Theses About Fictionality." *Narrative*, vol. 23, no. 1, 2015, pp. 61–73.

Niven-Phillips, Lisa. "Miss Piggy on Beauty." *British Vogue*, 31 Jan. 2012. https://www.vogue.co.uk/article/miss-piggy-beauty-interview-mac-collaboration.

Nussbaum, Abigail. "*Embassytown* by China Miéville." *Asking the Wrong Questions*, 27 Nov. 2011. https://wrongquestions.blogspot.com/2011/11/embassytown-by-china-mieville.html.

O'Brien, Flann. *At Swim-two-birds.* MacGibbon & Kee, 1961.

Oblivion. Directed by Joseph Kosinski, Universal Pictures, 2013.

Ortega, Julio. "Exchange System in *One Hundred Years of Solitude.*" *Gabriel García Márquez and the Powers of Fiction*, edited by Julio Ortega, assisted by Claudia Elliott, U of Texas P, 1988, pp. 1–16.

Ortuzar-Young, Ada. "Cuban Geographies: The Roots/Routes of Ana Menéndez Narratives." *Identity, Diaspora, and Return in American Literature*, edited by Maria Antònia Oliver Rotger, Routledge, 2014, pp. 149–69.

Osborne, Catherine. "Atomism." *The Oxford Classical Dictionary*, Oxford UP, 2015. https://doi.org/10.1093/acrefore/9780199381135.013.938.

"Override, *Tr.V.*" *American Heritage Dictionary of the English Language*, HarperCollins Publishers, 2022. https://ahdictionary.com/word/search.html?q=override. Accessed 28 Feb. 2025.

"Override, *Tr.V.*" *Collins English Dictionary*, HarperCollins Publishers. https://www.collinsdictionary.com/dictionary/english/override. Accessed 28 Feb. 2025.

"Override, *Tr. V.*" *Kernerman Webster's College Dictionary,* Random House, 2010. https://www.thefreedictionary.com/override. Accessed 28 Feb. 2025.

Paravisini-Gebert, Lizabeth. "Wicked Weeds: A Zombie Novel, by Pedro Cabiya." *Review: Literature and Arts of the Americas,* vol. 21, no. 2, 2018, pp. 304–6.

Parkkonen, Lauri, et al. "Early Visual Brain Areas Reflect the Percept of an Ambiguous Scene." *Proceedings of the National Academy of Sciences of the United States of America,* vol. 105, no. 51, 2008. https://doi.org/10.1073/pnas.0810966105.

Phelan, James. "The Mimetic, the Synthetic, and the Criterion of Correspondence; Or Audiences, the Mimetic Illusion, and Ghosts." *Debating Rhetorical Narratology: On the Synthetic, Mimetic, and Thematic Aspects of Narrative,* by Matthew Clark and James Phelan, Ohio State UP, 2020, pp. 151–70.

Phelan, James. *Narrative as Rhetoric: Technique, Audiences, Ethics, Ideology.* The Ohio State UP, 1996.

Pier, John. "Metalepsis." *Routledge Encyclopedia of Narrative Theory,* edited by David Herman et al., Routledge, 2008, pp. 303–4.

Plebe, Alessio. "Plant Sentience: Time Scale Matters." *Animal Sentience,* vol. 8, no. 33, 2023. https://doi.org/10.51291/2377-7478.1797.

Plotnitsky, Arkady. "Adventures of the Diagonal: Non-Euclidean Mathematics and Narrative." *Circles Disturbed: The Interplay of Mathematics and Narrative,* edited by Apostolos Doxiadis and Barry Mazur, Princeton UP, 2012, pp. 407–46.

Poe, Edgar Allan. "The Fall of the House of Usher." *Tales of the Grotesque and Arabesque,* vol. 1, Lea and Blanchard, 1840, pp. 75–103.

"Point of View in Photography: 4 Examples." *New York Film Academy,* 15 June 2023. https://www.nyfa.edu/student-resources/point-view-photography/.

Powers, Richard. *Orfeo.* W. W. Norton & Company, 2014.

Powers, Richard. *The Overstory.* W. W. Norton & Company, 2019.

Prince, Gerald. *Dictionary of Narratology.* Rev. ed., U of Nebraska P, 2003.

Purschke, Günter, et al. "Ultrastructure of Cerebral Eyes in Oweniidae and Chaetopteridae (Annelida)—Implications for the Evolution of Eyes in Annelida." *Zoological Letters,* vol. 8, no. 1, 25 Jan. 2022. https://doi.org/10.1186/s40851-022-00188-0.

"Quantum Entanglement." *A Dictionary of Physics,* edited by Jonathan Law and Richard Rennie, Oxford UP, 2019. https://www.oxfordreference.com/view/10.1093/acref/9780198821472.001.0001/acref-9780198821472-e-3737. Accessed 30 Dec. 2025.

Rabinowitz, Peter. *Before Reading: Narrative Conventions and the Politics of Interpretation.* The Ohio State UP, 1998.

Rabinowitz, Peter. "Truth in Fiction; A Reexamination of Audiences." *Critical Inquiry,* vol. 4, no. 1, 1977, pp. 121–41.

"Recombinant DNA." *Nature Education,* 2014. https://www.nature.com/scitable/definition/recombinant-dna-164/. Accessed 29 Dec. 2025.

Reid, Joe. "Today in TV History: 'The West Wing' Met 'Sesame Street.'" *Decider,* 3 Mar. 2016. https://decider.com/2016/03/03/today-in-tv-history-the-west-wing-met-sesame-street/.

Richards, I. A. *The Philosophy of Rhetoric.* Oxford UP, 1965.

Rodriguez, Kayla Cherice. *War and Play: Examining Julio Cortázar's War on Language in "Rayuela."* 2016. U of Colorado, master's thesis. *ProQuest* 10124057.

Rogers, Janine. *Unified Fields: Science and Literary Form.* McGill-Queen's UP, 2014.

Ryan, Marie-Laure. *Avatars of Story.* U of Minnesota P, 2006.

Saporta, Marc. *Composition No. 1*. Visual Editions, 2011.

Schank, Roger C., and Robert P. Abelson. "Knowledge and Memory: The Real Story." *Advances in Social Cognition*, vol. 8, 1995, pp. 1–85.

Schlosshauer, Maximilian. *Decoherence and the Quantum-To-Classical Transition*. Springer, 2008. Corrected third printing.

Schmerker, Jeff. "For These Species, It's All in the (Clonal) Family." *Integrated DNA Technologies*, 8 Jan. 2024. https://www.idtdna.com/pages/community/blog/post/for-these-species-it-s-all-in-the-(clonal)-family.

Schrödinger, Erwin. *What Is Life?* Cambridge UP, 1967.

Segundo-Ortin, Miguel, and Paco Calvo. "Plant Sentience? Between Romanticism and Denial: Science." *Animal Sentience*, vol. 8, no. 33, 2023. https://doi.org/10.51291/2377-7478.1772.

Semino, Elena. *Metaphor in Discourse*. Cambridge UP, 2008.

Shakespeare, William. *The Complete Works*. Edited by Stanley Wells, Gary Taylor, John Jowett, and William Montgomery, Clarendon Press, 2005.

Simard, Suzanne. *Finding the Mother Tree: Discovering the Wisdom of the Forest*. Knopf, 2021.

Simpkins, Scott. "'The Infinite Game': Cortázar's *Hopscotch*." *The Journal of the Midwest Modern Language Association*, vol. 23, no. 1, 1990, pp. 61–74.

Singh, Vandana. *Teaching Climate Change: Science, Stories, Justice*. Routledge, 2024.

Spengler, Birgit. "Arboreal Encounters in Richard Powers's *The Overstory*." *An Eclectic Bestiary: Encounters in a More-Than-Human World*, edited by Birgit Spengler and Babette B. Tischleder, Transcript Publishing, 2019, pp. 65–89.

Spiro, Rand J., and Jihn-Chang Jehng. "Cognitive Flexibility and Hypertext: Theory and Technology for the Nonlinear and Multidimensional Traversal of Complex Subject Matter." *Cognition, Education, and Multimedia: Exploring Ideas in High Technology*, edited by Don Nix and Rand J. Spiro, Lawrence Erlbaum, 1990, pp. 163–205.

Star, Susan Leigh, and James R. Griesemer. "Institutional Ecology, 'Translations' and Boundary Objects: Amateurs and Professionals in Berkeley's Museum of Vertebrate Zoology, 1907–39." *Social Studies of Science*, vol. 19, no. 3, 1989, pp. 387–420.

Steger, Cara, et al. "Ecosystem Services as Boundary Objects for Transdisciplinary Collaboration." *Ecological Economics*, vol. 143, 2018, pp. 153–60.

Stelle, Ginger. "'Starring Kermit the Frog as Bob Cratchit': Muppets as Actors." *Kermit Culture: Critical Perspectives on Jim Henson's Muppets*, edited by Jennifer C. Garlen and Anissa M. Graham, McFarland & Company, 2009.

Stoytcheva, Sveta. "A Field Guide: Part I: What Is a Boundary Object?" *Scalar*. https://scalar.usc.edu/works/boundary-objects-guide/boundary-objects.

Sullivan, Karen. *Frames and Constructions in Metaphoric Language*. John Benjamins, 2013.

Swift, Jonathan. *Gulliver's Travels*. Benjamin Motte, 1726.

"Syntax, *N*." *Online Etymology Dictionary*. https://www.etymonline.com/search?q=syntax. Accessed 3 Aug. 2025.

Thiher, Allen. *Words in Reflection: Modern Language Theory and Postmodern Fiction*. U of Chicago P, 1984.

Tobin, Patricia Drechsel. *Time and the Novel: The Genealogical Imperative*. Princeton UP, 1978.

Trauvitch, Rhona. "From Ancient Jewish Texts to Androids to AI, a Just-Right Sequence of Numbers or Letters Turns Matter into Meaning." *The Conversation*, 13 Oct. 2023. https://doi.org/10.64628/AAI.vhfks3a3q.

Trauvitch, Rhona. "Mapping with Fi-Sci: Why and How Fictionality Illuminates Science." *Narratologies of Science*, special issue of *Journal of Narrative Theory*, edited by Daniel Aureliano Newman, vol. 53, no. 1, 2023, pp. 59–86. https://doi.org/10.1353/jnt.2023.0000.

Trauvitch, Rhona. "Synontological Spaces." *The Routledge Handbook of Popular Culture and Tourism*, edited by Christine Lundberg and Vassilios Ziakas, Routledge, 2019.

"Verbal Analogies—Concepts for Aptitude." *LearnTheta*. https://www.learntheta.com/aptitude-verbal-analogies/. Accessed 4 Oct. 2023.

Verdié, Adriana. "Composing *Rayuela*: The Musical Element in Julio Cortázar's Narrative." *Australian Literary Studies*, vol. 29, nos. 1–2, 2014, pp. 130–47.

Vonnegut, Kurt. *Slaughterhouse-Five*. Delacorte, 1969.

Wallace, David Foster. *Infinite Jest*. Little, Brown, 1996.

Walsh, Richard. *The Rhetoric of Fictionality: Narrative Theory and the Idea of Fiction*. The Ohio State UP, 2007.

Waugh, Patricia. *Metafiction: The Theory and Practice of Self-Conscious Fiction*. Methuen, 1984.

"The Wendy Williams Show—Interview with Miss Piggy." *YouTube*, uploaded by HowieKemperVideo, 13 Jan. 2015. https://www.youtube.com/watch?v=lTJ4pTzTYgw.

"What Are the Gestalt Principles?" *Interaction Design Foundation—IxDF*, 30 Aug. 2016. https://www.interaction-design.org/literature/topics/gestalt-principles.

"What Is Epigenetics, and What Does It Mean for CIPHERS?" *CIPHERS*, 6 June 2019. https://sites.duke.edu/ciphers/2019/06/06/part-2-what-is-epigenetics-and-what-does-it-mean-for-ciphers/.

"What Is Pando." *Friends of Pando*. https://www.friendsofpando.org/what-is-pando/. Accessed 1 Jan. 2026.

"What 'Motherboard' Really Means—And Why It Matters." *YouTube*, uploaded by History of Simple Things, 12 Oct. 2025. https://www.youtube.com/watch?v=oLuociFFHa8.

"When Knowledge Conquered Fear." *Cosmos: A Spacetime Odyssey*, season 1, episode 3, National Geographic, 23 Mar. 2014.

Wilson, Alan. *Knowledge Power: Interdisciplinary Education for a Complex World*. Routledge, 2010.

Wilson, Alan. "Routes to Interdisciplinarity 1: Ten Superconcepts." *The London Interdisciplinary School*, 9 Jan. 2024. https://www.lis.ac.uk/stories/routes-to-interdisciplinarity-ten-superconcepts.

Winterson, Jeanette. *Gut Symmetries*. Knopf, 1997.

Woo, Marcus. "Will We Ever . . . Travel in Wormholes?" *BBC*, 25 Mar. 2014. https://www.bbc.com/future/article/20140326-will-we-ever-travel-in-wormholes.

Wood, Michael. *Gabriel García Márquez: One Hundred Years of Solitude*. Cambridge UP, 1990.

"The World Is Poorly Designed. But Copying Nature Helps." *YouTube*, uploaded by Vox, 3 Nov. 2017. https://www.youtube.com/watch?v=iMtXqTmftao.

"Wormhole." *A Dictionary of Physics*. Edited by Jonathan Law and Richard Rennie, Oxford UP, 2019. https://www.oxfordreference.com/view/10.1093/acref/9780198714743.001.0001/acref-9780198714743-e-3311. Accessed 29 Aug. 2025.

Zeh, H. Dieter. "On the Interpretation of Measurement in Quantum Theory." *Foundations of Physics*, vol. 1, no. 1, 1970, pp. 69–76.

Zurek, Wojciech H. "Decoherence and the Transition from Quantum to Classical—Revisited." *Los Alamos Science*, no. 27, 2002, pp. 2–25.

INDEX

THEORY AND INTERPRETATION OF NARRATIVE

James Phelan, Katra Byram, and Faye Halpern, Series Editors
Robyn Warhol and Peter Rabinowitz, Founding Editors Emeriti

Because the series editors believe that the most significant work in narrative studies today contributes both to our knowledge of specific narratives and to our understanding of narrative in general, studies in the series typically offer interpretations of individual narratives and address significant theoretical issues underlying those interpretations. The series does not privilege one critical perspective but is open to work from any strong theoretical position.

Fi-Sci: Avatars of Science in Fiction by Rhona Trauvitch

Narrative Co-Construction: Author-Audience Interactions and Their Contexts edited by Malcah Effron, Margarida McMurry, and Virginia Pignagnoli

Disrupting Narrative Rhetoric: Experimental Fictionality on Digital Platforms by Stefan Iversen

Contemporary Feminist Fiction and a Case for Expanding Rhetorical Narratology by Katherine J. Weese

Never on Time, Always in Time: Narrative Form and the Queer Sensorium by Kate McCullough

Slow Narrative across Media edited by Marco Caracciolo and Ella Mingazova

Deformative Fictions: Cruelty and Narrative Ethics in Twentieth-Century Latin American Literature by Ashley Hope Pérez

Post-Postmodernist Fiction and the Rise of Digital Epitexts by Virginia Pignagnoli

Becoming Pynchon: Genetic Narratology and "V." by Luc Herman and John M. Krafft

The Story of Fictional Truth: Realism from the Death to the Rise of the Novel by Paul Dawson

Fictionality and Literature: Core Concepts Revisited edited by Lasse R. Gammelgaard, Stefan Iversen, Louise Brix Jacobsen, James Phelan, Richard Walsh, Henrik Zetterberg-Nielsen, and Simona Zetterberg-Nielsen

A New Anatomy of Storyworlds: What Is, What If, As If by Marie-Laure Ryan

Narrative in the Anthropocene by Erin James

Experiencing Visual Storyworlds: Focalization in Comics by Silke Horstkotte and Nancy Pedri

With Bodies: Narrative Theory and Embodied Cognition by Marco Caracciolo and Karin Kukkonen

Digital Fiction and the Unnatural: Transmedial Narrative Theory, Method, and Analysis by Astrid Ensslin and Alice Bell

Narrative Bonds: Multiple Narrators in the Victorian Novel by Alexandra Valint

Contemporary French and Francophone Narratology edited by John Pier

We-Narratives: Collective Storytelling in Contemporary Fiction by Natalya Bekhta

Debating Rhetorical Narratology: On the Synthetic, Mimetic, and Thematic Aspects of Narrative by Matthew Clark and James Phelan

Environment and Narrative: New Directions in Econarratology edited by Erin James and Eric Morel

Unnatural Narratology: Extensions, Revisions, and Challenges edited by Jan Alber and Brian Richardson

A Poetics of Plot for the Twenty-First Century: Theorizing Unruly Narratives by Brian Richardson

Playing at Narratology: Digital Media as Narrative Theory by Daniel Punday

Making Conversation in Modernist Fiction by Elizabeth Alsop

Narratology and Ideology: Negotiating Context, Form, and Theory in

Postcolonial Narratives edited by Divya Dwivedi, Henrik Skov Nielsen, and Richard Walsh

Novelization: From Film to Novel by Jan Baetens

Reading Conrad by J. Hillis Miller, Edited by John G. Peters and Jakob Lothe

Narrative, Race, and Ethnicity in the United States edited by James J. Donahue, Jennifer Ann Ho, and Shaun Morgan

Somebody Telling Somebody Else: A Rhetorical Poetics of Narrative by James Phelan

Media of Serial Narrative edited by Frank Kelleter

Suture and Narrative: Deep Intersubjectivity in Fiction and Film by George Butte

The Writer in the Well: On Misreading and Rewriting Literature by Gary Weissman

Narrating Space / Spatializing Narrative: Where Narrative Theory and Geography Meet by Marie-Laure Ryan, Kenneth Foote, and Maoz Azaryahu

Narrative Sequence in Contemporary Narratology edited by Raphaël Baroni and Françoise Revaz

The Submerged Plot and the Mother's Pleasure from Jane Austen to Arundhati Roy by Kelly A. Marsh

Narrative Theory Unbound: Queer and Feminist Interventions edited by Robyn Warhol and Susan S. Lanser

Unnatural Narrative: Theory, History, and Practice by Brian Richardson

Ethics and the Dynamic Observer Narrator: Reckoning with Past and Present in German Literature by Katra A. Byram

Narrative Paths: African Travel in Modern Fiction and Nonfiction by Kai Mikkonen

The Reader as Peeping Tom: Nonreciprocal Gazing in Narrative Fiction and Film by Jeremy Hawthorn

Thomas Hardy's Brains: Psychology, Neurology, and Hardy's Imagination by Suzanne Keen

The Return of the Omniscient Narrator: Authorship and Authority in Twenty-First Century Fiction by Paul Dawson

Feminist Narrative Ethics: Tacit Persuasion in Modernist Form by Katherine Saunders Nash

Real Mysteries: Narrative and the Unknowable by H. Porter Abbott

A Poetics of Unnatural Narrative edited by Jan Alber, Henrik Skov Nielsen, and Brian Richardson

Narrative Discourse: Authors and Narrators in Literature, Film, and Art by Patrick Colm Hogan

An Aesthetics of Narrative Performance: Transnational Theater, Literature, and Film in Contemporary Germany by Claudia Breger

Literary Identification from Charlotte Brontë to Tsitsi Dangarembga by Laura Green

Narrative Theory: Core Concepts and Critical Debates by David Herman, James Phelan and Peter J. Rabinowitz, Brian Richardson, and Robyn Warhol

After Testimony: The Ethics and Aesthetics of Holocaust Narrative for the Future edited by Jakob Lothe, Susan Rubin Suleiman, and James Phelan

The Vitality of Allegory: Figural Narrative in Modern and Contemporary Fiction by Gary Johnson

Narrative Middles: Navigating the Nineteenth-Century British Novel edited by Caroline Levine and Mario Ortiz-Robles

Fact, Fiction, and Form: Selected Essays by Ralph W. Rader edited by James Phelan and David H. Richter

The Real, the True, and the Told: Postmodern Historical Narrative and the Ethics of Representation by Eric L. Berlatsky

Franz Kafka: Narration, Rhetoric, and Reading edited by Jakob Lothe, Beatrice Sandberg, and Ronald Speirs

Social Minds in the Novel by Alan Palmer

Narrative Structures and the Language of the Self by Matthew Clark

Imagining Minds: The Neuro-Aesthetics of Austen, Eliot, and Hardy by Kay Young

Postclassical Narratology: Approaches and Analyses edited by Jan Alber and Monika Fludernik

Techniques for Living: Fiction and Theory in the Work of Christine Brooke-Rose by Karen R. Lawrence

Towards the Ethics of Form in Fiction: Narratives of Cultural Remission by Leona Toker

Tabloid, Inc.: Crimes, Newspapers, Narratives by V. Penelope Pelizzon and Nancy M. West

Narrative Means, Lyric Ends: Temporality in the Nineteenth-Century British Long Poem by Monique R. Morgan

Understanding Nationalism: On Narrative, Cognitive Science, and Identity by Patrick Colm Hogan

Joseph Conrad: Voice, Sequence, History, Genre edited by Jakob Lothe, Jeremy Hawthorn, James Phelan

The Rhetoric of Fictionality: Narrative Theory and the Idea of Fiction by Richard Walsh

Experiencing Fiction: Judgments, Progressions, and the Rhetorical Theory of Narrative by James Phelan

Unnatural Voices: Extreme Narration in Modern and Contemporary Fiction by Brian Richardson

Narrative Causalities by Emma Kafalenos

Why We Read Fiction: Theory of Mind and the Novel by Lisa Zunshine

I Know That You Know That I Know: Narrating Subjects from "Moll Flanders" to "Marnie" by George Butte

Bloodscripts: Writing the Violent Subject by Elana Gomel

Surprised by Shame: Dostoevsky's Liars and Narrative Exposure by Deborah A. Martinsen

Having a Good Cry: Effeminate Feelings and Pop-Culture Forms by Robyn R. Warhol

Politics, Persuasion, and Pragmatism: A Rhetoric of Feminist Utopian Fiction by Ellen Peel

Telling Tales: Gender and Narrative Form in Victorian Literature and Culture by Elizabeth Langland

Narrative Dynamics: Essays on Time, Plot, Closure, and Frames edited by Brian Richardson

Breaking the Frame: Metalepsis and the Construction of the Subject by Debra Malina

Invisible Author: Last Essays by Christine Brooke-Rose

Ordinary Pleasures: Couples, Conversation, and Comedy by Kay Young

Narratologies: New Perspectives on Narrative Analysis edited by David Herman

Before Reading: Narrative Conventions and the Politics of Interpretation by Peter J. Rabinowitz

Matters of Fact: Reading Nonfiction over the Edge by Daniel W. Lehman

The Progress of Romance: Literary Historiography and the Gothic Novel by David H. Richter

A Glance Beyond Doubt: Narration, Representation, Subjectivity by Shlomith Rimmon-Kenan

Narrative as Rhetoric: Technique, Audiences, Ethics, Ideology by James Phelan

Misreading "Jane Eyre": A Postformalist Paradigm by Jerome Beaty

Psychological Politics of the American Dream: The Commodification of Subjectivity in Twentieth-Century American Literature by Lois Tyson

Understanding Narrative edited by James Phelan and Peter J. Rabinowitz

Framing Anna Karenina: Tolstoy, the Woman Question, and the Victorian Novel by Amy Mandelker

Gendered Interventions: Narrative Discourse in the Victorian Novel by Robyn R. Warhol

Reading People, Reading Plots: Character, Progression, and the Interpretation of Narrative by James Phelan